Vectors and Matrices for Geometric and 3D Modeling

Michael E. Mortenson does independent research and writes on topics in geometric and 3D modeling. He is a former research scientist with a major aerospace corporation and the author of several successful textbooks, including *Geometric Modeling, Third Edition, Mathematics for Computer Graphics Applications,* Second Edition, *Geometric Transformations for 3D Modeling,* Second Edition, and *3D Modeling, Animation, and Rendering: An Illustrated Lexicon.* These works have garnered over 2400 citations. He is a graduate of the UCLA School of Engineering.

Vectors and Matrices for Geometric and 3D Modeling

Michael E. Mortenson

INDUSTRIAL PRESS, INC.

Industrial Press, Inc.
32 Haviland Street, Suite 3
South Norwalk, Connecticut 06854
Phone: 203-956-5593
Toll-Free in USA: 888-528-7852
Email: info@industrialpress.com

Author: Michael E. Mortenson
Title: Vectors and Matrices for Geometric and 3D Modeling
Library of Congress Control Number: 2020933436

Printed in the United States of America.

ISBN (print) 978-0-8311-3655-0
ISBN (ePUB) 978-0-8311-9562-5
ISBN (eMOBI) 978-0-8311-9563-2
ISBN (ePDF) 978-0-8311-9561-8

Publisher/Editorial Director: Judy Bass
Copy Editor: Judy Duguid
Compositor: Paradigm Data Services (P) Ltd., Chandigarh
Proofreader: Claire Splan
Indexer: Claire Splan

books.industrialpress.com
ebooks.industrialpress.com

10 9 8 7 6 5 4 3 2 1

To
JAM

Aloha Nui Loa

Contents

Acknowledgments

Over sixty years ago, many talented men and women in both academia and industry succeeded in bringing the theories and principles of higher mathematics, such as differential geometry and linear algebra, into the practical world of engineering design and manufacturing. (Computational geometry began to mature in those early years, too.) These researchers took advantage of the growing power of the digital computer and the potential of a graphic user interface to tame the difficult mathematics and bring it to the desktop and drafting board. The application of vectors and matrices to geometric and 3D modeling owes its success to these researchers. Many thanks to all of them.

Thanks to the very talented staff at Industrial Press, Inc. for their support and guidance. I especially thank Judy Bass, Publisher, who unerringly kept this project moving each step of the way. Also, I thank Patricia Wallenberg, Devenand Madhuhar and their staff as Compositor, and Judy Duguid, Copyeditor. (Ms. Duguid, amazingly, could spot small inconsistencies separated by dozens of pages.)

Special thanks to my wife, Janet. Her proofreading skills once again saved me from revealing a large portion of my grammatical shortcomings before I sent the final draft of this book to the publisher. I dare not remind her here how many books she has given her time to.

Finally, credit Norton Juster, author of *The Dot and the Line*, for part of the quote in the Preface, "To the vector belong the spoils," and thanks to Rena and Earll Murman, good friends and neighbors, for bringing it to my attention. As far as I can tell, the aphorism "to the victor go the spoils" was first uttered by Senator Wm. L. Marcy, 1831. But one can easily imagine one of the Caesars declaring it much earlier.

Michael E. Mortenson
Port Townsend, Washington
2020

Preface

To the vector belong the spoils,
To the matrix belong the toils.*

Vectors and Matrices for Geometric and 3D Modeling introduces the mathematics that is the foundation of computer graphics applications. Chapters are organized from introductory to more complex topics. Within each chapter, there is a similar order of elementary-to-advanced discussion. Topics are covered that usually appear in briefer form in more advanced textbooks on modeling and only as a small part of their supporting mathematics. In this textbook, vectors and matrices are the main subjects. Basic concepts are presented in an easy-to-understand and thoroughly illustrated form, without the burden of formalisms. This leads directly to how vectors and matrices are used in geometric and 3D modeling.

The relationships between vectors, matrices, basis vectors, and barycentric coordinates are not usually seen together in an elementary textbook. How these concepts are applied to produce curves and surfaces and how they aid in analysis of spatial relationships are discussed. The mathematics of vectors and matrices extends to CAD/CAM, animation, and rendering, so this textbook may serve as a supplement to introductory courses on these subjects.

The level of discussion in this book assumes an understanding of high school geometry and algebra, including some algebraic geometry and simple linear equations. A rigorous introduction to vectors and matrices usually happens in linear algebra courses. But linear algebra is not required to understand the concepts presented here. With this background the reader will easily progress through most of this textbook and, with some additional effort, through the brief discussions of more advanced math.

Vectors are perhaps the most important mathematical objects used in modeling and animation. They have the properties of magnitude and direction, represented pictorially by a line segment with an arrowhead. With these and other characteristics, vectors provide intuitive, often visual, understanding of model construction and analysis.

---

* See Acknowledgments.

Vector algebra offers distinct advantages over traditional algebraic geometry. For example, vector functions are able to define curves and surfaces that twist in space and that are easily modified. Vector algebra offers efficient solutions to complex geometric problems, and can minimize computational dependence on a specific coordinate system until the later stages of modeling and problem solving.

Matrices are natural and hardworking partners of vectors. A matrix is a set of numbers or other mathematical elements arranged in a rectangular array of rows and columns. Matrix algebra is an efficient way to represent and manipulate vectors. The structure of a matrix makes it easy to assemble and to work with mathematical data. For example, we are able to organize as elements of a matrix the coefficients of a set of simultaneous linear equations, the coordinates of a point, the components of a vector, or the components of a geometric transformation matrix.

Chapter 1 gives a brief overview of vectors and their mathematical cousins. Eleven more chapters delve into the details of vectors and matrices, beginning with the most basic concepts and progressing to discussions of how they are related to geometric and 3D modeling. Some sections of the last three chapters call for a minimum familiarity with beginning calculus on the part of the reader. For a presentation with more formal mathematics, readers should refer to advanced textbooks on these subjects. For practice in understanding the concepts, some of the chapters end with exercises. Answers to all of them appear at the end of the book, some with comments on how the solution is achieved. Altogether there are over 150 illustrations to reinforce the text, and there are over 100 exercises.

Those readers familiar with the author's *Geometric Modeling* and *Geometric Transformations for 3D Modeling* (both published by Industrial Press) will find some topics and illustrations repeated here, with less formality as is suitable for an entry-level textbook such as this one. For many readers these two textbooks will be a natural follow-on to more advanced discussions.

So who should read this book? For those readers beginning studies in geometric and 3D modeling, animation, CGI, or CAD/CAM, this book serves as an introduction to vectors and matrices and provides a good start on understanding how they are applied. For an instructor, this book serves as a primary text or as a supplement to more advanced or specialized texts on geometric and 3D modeling. Finally, for those now working in these fields, this textbook provides another way to look at vectors and matrices, as well as serves as a helpful review and reference.

Michael E. Mortenson
Port Townsend, Washington
2020

Vectors and Matrices for Geometric and 3D Modeling

1 Vectors PDQ

Vectors are among the most important mathematical tools that help us define geometric and 3D models. They also let us operate on these models to change their shape and analyze their properties. These operations include rigid-body transformations, deformations, projections, and analysis of spatial relationships. We can define vectors in the familiar orthogonal three-dimensional Euclidean space or in nonorthogonal spaces, defined by basis vectors.

Warning to the reader: *This chapter is a broad overview of the characteristics of a vector and its mathematical cousins. It rapidly covers concepts that some of you may not be familiar with. The chapters that follow will explore in depth the ideas merely summarized here.*

1.1 A Very Brief Overview

1. A vector is a mathematical structure derived and built from the materials of linear algebra.
2. A vector has the properties of magnitude or length and direction. Imagine an arrowheaded line segment.
3. A vector is represented by a set of components; in three-dimensional space it has three components, each lying along a different coordinate axis or basis vector.
4. A vector is a geometric object that operates on itself and other geometric objects to produce new vectors or geometric objects.
5. A vector operating on itself or another vector via the inner product produces two scalar quantities that are unchanged under certain coordinate system transformations: a magnitude or length and an angle.
6. A vector or set of vectors can represent more complex geometric objects and their local and global properties.
7. A vector is a member of an extended family of mathematical structures that includes tensors, matrices, scalars, and their hybrid cousins . . . quaternions.

1.2 With Comments Added

Each of the above seven statements sounds like a definition. And if so, we immediately see that a vector has many definitions. None are adequate in themselves, so we must proceed with some vagueness. But by the end of this work, you will know what a vector is, just not how to define it in a single short statement. Here are some comments on each of those statements:

1. *A vector is a mathematical structure derived and built from the materials of linear algebra.*

   This is a somewhat abstract definition. However, linear algebra is where many students first learn about vectors and their relationship to linear equations. Linear algebra is the mathematics dealing with equations like this:

   $$ax+by+cz=d, \tag{1.1}$$

   or more generally,

   $$a_1x_1+a_2x_2+\cdots+a_nx_n=d. \tag{1.2}$$

   You are probably more familiar with using the tools of linear algebra to solve simple systems of linear equations, such as:

   $$\begin{aligned} ax+by&=c\\ dx+ey&=f \end{aligned} \tag{1.3}$$

   That is, find the values of x and y that satisfy the two equations. Okay, this is a pretty simple set of equations. But you get the idea.

   It turns out that the mathematic and geometric characteristics of a vector arose independently and before today's theories of linear algebra. The theory of vectors found a good fit within linear algebra, as did its matrix and determinant cousins. And all of these have a relationship with tensors.

   As for the phrase "mathematical structures," aside from the sense of building something from something else, formally it has a more complex meaning, which we have no need to explore here.

2. *A vector has the properties of magnitude or length and direction.*

   This is the more popular, intuitive definition. It says that we can draw a picture of a vector: We can visualize a vector as an arrow, that is, an arrowheaded line segment. Although this is a good start toward understanding vectors, it doesn't go far enough.

   Magnitude is a scalar, and direction is like another vector, a unit vector pointing in a specific direction. (This sounds a bit circular, but later discussion will resolve this impression.) Performing the inner product on a vector with itself produces its magnitude. A vector's components (or its unit vector components) indicate its direction; as it turns out, they (the unit vector's components) are its direction cosines in the familiar Cartesian coordinate system. For example, a vector is a good way to define a translation, specifying how far and in what direction. Chapters 2 and 3 explain these characteristics in detail.

3. *A vector is represented by a set of components; in three-dimensional space it has three components, each lying along a different coordinate axis or basis vector.*

   The components are also vectors. They are collinear with the basis vectors that define the coordinate system. If the coordinate system changes (that is, if the basis vectors change), the components change as well. So a vector's components depend on the coordinate system in which the vector finds itself and its orientation within it. Chapter 6 is all about basis vectors and coordinate systems.

4. *A vector is a geometric object that operates on itself and other geometric objects to produce new vectors or geometric objects.*

   Points, lines, planes, polygons, polyhedra, curves, surfaces, solids, vectors, and tensors (which, in a way, generalize vectors, so they should be included) are examples of what it means to be a geometric object. Vector operations include the inner product (also known as the scalar or dot product), cross product (also known as the vector product), and many more. See Chapter 3 for more examples.

5. *A vector operating on itself or another vector via the inner product produces two scalar quantities that are unchanged under certain coordinate system transformations: a magnitude or length and an angle.*

   There are several ways to write the inner product of a vector with itself. Here are some of them:

   a. $\mathbf{p} \bullet \mathbf{p} = |\mathbf{p}|^2$

   b. $\mathbf{p} \bullet \mathbf{p} = \|\mathbf{p}\|^2$

   c. $\mathbf{p} \bullet \mathbf{p} = p_x p_x + p_y p_y + p_z p_z$

   d. $\mathbf{p} \bullet \mathbf{p} = \langle \mathrm{p}, \mathrm{p} \rangle$

Bold lowercase letters indicate vectors. On the left side of the four equations above is the fairly common dot (inner) product notation. The notation on the right side of the last equation is more common in physics, especially quantum physics. More discussion of the inner product appears in Chapter 3.

And finally, what are "certain coordinate system transformations"? Simple translation and rotation of a coordinate system to produce a new, transformed, coordinate system do not change the magnitude of the inner product of two vectors. More about this also in Chapter 3.

6. *A vector or set of vectors can represent more complex geometric objects and their local and global properties.*

   See the list of geometric objects in item 4, above. As for properties, vectors may represent tangents and normals to curves and surfaces, as well as rotations, torsions, the curvature vector, and much else.

7. *A vector is a member of an extended family of mathematical structures that includes tensors, matrices, scalars, and their hybrid cousins ... quaternions.*

   Tensors are a generalization of vectors. Or the other way around, vectors and scalars are special kinds of tensors. A vector has a single subscript or superscript index identifying its components (more on indexing in Section 1.4 below). A tensor has one or more indices. The matrices we will work with are usually one- or two-dimensional arrays but can be n-dimensional arrays. Row or column matrices often represent a vector. So, if **P** is a matrix containing the coordinates of a point, there are several ways to write it. For example, as row matrices:

   $$\mathrm{P}=[p_x \ \ p_y \ \ p_z], \quad \mathrm{P}=[x \ \ y \ \ z],$$

   or as column matrices:

   $$\mathrm{P}=\begin{bmatrix} p_x \\ p_y \\ p_z \end{bmatrix}, \qquad \mathrm{P}=\begin{bmatrix} x \\ y \\ z \end{bmatrix}.$$

   Tensors describe linear relationships between vectors, scalars, and also other tensors. Matrices help organize all of this, and they come with special rules describing how they operate to perform transformations, to rearrange, to simplify, to solve, and more.

1.3 And What Follows

Eleven more chapters follow this one. Each is focused on a group of concepts and operations using vectors and matrices that have a track record of usefulness to geometric and 3D modeling and more, as these fields move into the era of artificial intelligence and quantum computing.

The study of vectors often begins with a discussion of their geometric and visual interpretation. This is what Chapter 2 does. It shows that simple geometric constructions reveal how to decompose a vector into its components, which, when added appropriately, reconstruct the original vector. It shows how to add vectors and more.

Chapter 3 builds upon and goes beyond the visual geometric interpretation. It describes the more powerful algebraic description of vectors. It reveals their more abstract qualities, although often with a geometric or visual side. Here is where you will begin to see the computational value of vectors in geometric modeling.

Chapters 4 and 5 focus on matrices: the basics (Chapter 4) and the special role of square matrices (Chapter 5), including eigenvectors, eigenvalues, and determinants.

Basis vectors and coordinate systems are the focus of Chapter 6. Basis vectors define the coordinate systems inhabited by vectors. They establish the familiar orthonormal Cartesian system in Euclidean space, oblique systems, and more general systems, where the distance scale may be different along each of the three coordinate axes. Add a locating point, not necessarily the origin, to a set of basis vectors, and you have a frame. These tools are enormously helpful.

Barycentric coordinate systems are unusual: They do not have an origin, principal axes, or principal planes . . . at least not in the more familiar sense. But a barycentric coordinate system is good at locating points within a triangle, which has application to shading and rendering the polygonal faces of 3D models for digital graphic display. Chapter 7 presents the mathematics of these systems. And, of course, vectors play a significant role.

Chapter 8 is about the rigid-body geometric transformations of translation and rotation. Translation is as simple as adding vectors, while there are many ways to produce a rotation. The chapter closes with a brief look at kinematics, which relies heavily on rigid-body transformations.

Chapter 9 is about more geometric transformations, including more about translation and rotation, scaling, shear, and projection transformations. It introduces the idea of combining transformations, using homogeneous coordinates.

Chapters 10 and 11 show how to use vectors to define basic geometric elements. The simplest of these elements are points, lines, and planes. These chapters show how to go from the parametric equations of curves and surfaces to their equivalent vector representations.

Chapter 12 is all about using vectors to reveal spatial relationships between geometric objects. It discusses distances and intersections.

There are over 100 exercises, mainly dealing with vector algebra. A solution section following Chapter 12 presents answers and comments to most of the exercises.

1.4 Backstory: The Very Short Version

Some historians of vectors claim to trace the origin of vectors at least back to Euclid. The ultimate roots of all mathematics go back to counting and drawing figures in the sand or on clay tablets. With the invention of early Cartesian analytic geometry and complex numbers, it became apparent that an object that had direction and magnitude was possible. That object became the vector of engineering and physics.

Sometime later, mathematicians, those of a more theoretical bent, found that they could derive the idea of a vector from the principles of linear algebra. They defined a vector as an ordered set of numbers, such that the vector **a** is represented by $(a_1, a_2, \ldots, a_n)$. So that if $n = 3$, for example, then **a** is a three-dimensional vector, having, of course, three components. And if they assigned certain rules establishing how vectors operate on each other, then they could extract new vectors and geometric information about them and about the vector space that is produced. One rule might be that if two vectors are added, $\mathbf{a} + \mathbf{b}$, then a new vector is produced, $\mathbf{c} = \mathbf{a} + \mathbf{b}$, if and only if $c_1 = a_1 + b_1$, $c_2 = a_2 + b_2, \ldots, c_n = a_n + b_n$. Other rules govern how vectors are multiplied, by either scalars or other vectors. These concepts and rules define a linear vector space.

So this textbook begins with an introduction to the visual, arrow-headed vector, proceeds to its algebraic interpretation, and then briefly offers some further grounding in linear vector spaces.

1.5 Notation . . . Motivation

The notation forms for vectors and matrices are legion. This section reviews vector and matrix notation and the use of indexing. Indexing is a shorthand way of keeping track of terms and controlling operations and not losing one's way in a repetitious thicket of constants and variables. Proper and consistent notation and indexing convey a lot of information and are indispensable in forming, in the simplest way, mathematical statements and operations. Think of notation as a way to simplify, clarify, and control. This is a section to return to for reference when in doubt about the meaning or use of a symbol.

Vectors

Here are some of the more common ways to indicate a vector, its operations, and its close mathematical relatives:

$a, b, c \ldots$	Scalars, constant coefficients
$\mathbf{a}, \mathbf{b}, \mathbf{p}, \mathbf{r}, \mathbf{s}, \ldots$	Vectors: boldface, lowercase letters
$\mathbf{i}, \mathbf{j}, \mathbf{k}$	Orthogonal unit basis vectors
$\lvert\mathbf{a}\rvert$ or $\lVert\mathbf{a}\rVert$ or a	Magnitude, length of **a**, also norm
$\hat{\mathbf{a}}$	Unit vector: $\hat{\mathbf{a}} = \mathbf{a}/\lvert\mathbf{a}\rvert$
$\mathbf{a} + \mathbf{b}$	Vector addition
$k\mathbf{a}$	Vector multiplied by a scalar

$\mathbf{a} \cdot \mathbf{a}$ or $\mathbf{a} \cdot \mathbf{b}$	Scalar or inner product
$\mathbf{a} \times \mathbf{b}$	Vector product of two vectors
$\mathbf{a} \cdot (\mathbf{b} \times \mathbf{c})$	Scalar-valued triple product
$\mathbf{a} \times (\mathbf{b} \times \mathbf{c})$	Vector-valued triple product
$\mathbf{e}_1, \mathbf{e}_2, \mathbf{e}_3, \ldots, \mathbf{e}_n$	Basis vectors, n-dimensional space
$e_1, e_2, e_3, \ldots, e_n$	Basis vector components
$\mathbf{p} = (p_x, p_y, p_z)$	Ordered set of vector components
$\mathbf{p} = p_x\mathbf{i} + p_y\mathbf{j} + p_z\mathbf{k}$	Vector as sum of its components
$\mathrm{P} = [p_x \; p_y \; p_z]$	Row matrix (vector components)
$\mathrm{P} = \begin{bmatrix} p_x \\ p_y \\ p_z \end{bmatrix}$	Column matrix (vector components)
$\mathbf{p}(u)$	Vector function of u
$\mathbf{p}(u, w)$	Vector function of u and w
$\mathbf{p}(u_i)$ or $\mathbf{p}_i$	Vector function evaluated at $u = u_i$
$\mathbf{p}(0)$ or $\mathbf{p}_0$	Vector function evaluated at $u = 0$
$\mathbf{p}(1)$ or $\mathbf{p}_1$	Vector function evaluated at $u = 1$
$\mathbf{p}(u_i, w_j)$	Vector function evaluated at $u = u_i$ and $w = w_j$
$d\mathbf{p}(u)/du$ or $\mathbf{p}^u$	First derivative of the function $\mathbf{p}(u)$ (Note that the superscript u, on $\mathbf{p}^u$, is not an exponent. It indicates differentiation with respect to the independent parametric variable u.)
$\bar{v}, \vec{v}, \bar{v}$	Handwritten vector notation
$Q = a + bi + cj + dk$	Quaternion
i, j, k	Quaternion units as complex numbers: $i^2, j^2, k^2 = -1$
$i = j = k = \sqrt{-1}$	Quaternion units definition
$i^2 = j^2 = k^2 = ijk = -1$	Quaternion units relationships
$ij = k \quad ji = -k$ $jk = i \quad kj = -i$ $ki = j \quad ik = -j$	More quaternion relationships
$Q = a + \mathbf{q}$	Quaternion as a scalar plus a vector

Matrices

Matrices organize the elements of vectors and algebraic data and expedite the transformations that shape and position geometric objects. Here are some of the ways they appear and are used:

$\mathbf{A}, \mathbf{B}, \mathbf{M}, \ldots$	Matrices: boldface, uppercase letters
$\mathbf{A} = \begin{bmatrix} a_{11} & a_{12} & a_{13} \\ a_{21} & a_{22} & a_{23} \\ a_{31} & a_{32} & a_{33} \end{bmatrix}$	3x3 square matrix
$a_{ij}, b_{ij}, m_{ij} \ldots$	Matrix elements
$m \times n$	Matrix size (m rows, n columns)
$\mathrm{A} = [a_{11}\ a_{12}\ a_{13}]$	Row matrix
$\mathrm{A} = \begin{bmatrix} a_{11} \\ a_{21} \\ a_{31} \end{bmatrix}$	Column matrix
$\mathbf{A} = \begin{bmatrix} a_{11} & 0 & 0 \\ 0 & a_{22} & 0 \\ 0 & 0 & a_{33} \end{bmatrix}$	Diagonal matrix (square matrices only)
$\mathbf{A} + \mathbf{B}$	Matrix addition
$\mathbf{AB}$	Matrix multiplication
$\mathbf{I} = \begin{bmatrix} 1 & 0 & 0 \\ 0 & 1 & 0 \\ 0 & 0 & 1 \end{bmatrix}$	3x3 identity matrix
$\delta_{ij} = \begin{cases} 0 \text{ if } i \neq j \\ 1 \text{ if } i = j \end{cases}$	Kronecker delta
$\delta_{ij} = \mathbf{I}_{ij}$	Equivalent in linear algebra
$\varepsilon_{ij}, \varepsilon_{ijk}, \varepsilon_{ijkl}$	Levi-Civita symbol
$\varepsilon_{ijk\ldots} = 0$	If any two indices are equal
$\varepsilon_{ijk\ldots} = +1$	If the indices are in cyclic order
$\varepsilon_{ijk\ldots} = -1$	If the indices are in anticyclic order
If $\mathbf{A}^{\mathrm{T}}\mathbf{A} = \mathbf{A}\mathbf{A}^{\mathrm{T}} = \mathbf{I}$	Then **A** is an orthogonal matrix
$\mathbf{A}^{-1},\ \mathbf{B}^{-1},\ \mathbf{M}^{-1}$	Matrix inverse

$\mathbf{A}^{\mathrm{T}}, \mathbf{B}^{\mathrm{T}}, \mathbf{M}^{\mathrm{T}}$	Matrix transpose
$\mathbf{p}', \mathbf{p}^*, \mathbf{M}', \mathbf{M}^*$	Transformed vector or matrix
$\lvert\mathbf{A}\rvert$ or $\det\mathbf{A}$	Determinant; a square matrix

Indexing

Indexing is the use of subscripts and superscripts to identify vector, matrix, and tensor elements and to indicate rules of combining them.

a_i	Subscript i
a^i	Superscript i
a_{ij}	Double subscripts i and j
a^{ij}	Double superscripts i and j
a^i_{jk}	Multiple subscripts and superscripts
$\sum_{i=1}^{4} a_i = a_1 + a_2 + a_3 + a_4$	Summation indexing
$x^i \mathbf{e}_i$	Einstein summation convention
$x^i \mathbf{e}_i = x^1 \mathbf{e}_1 + x^2 \mathbf{e}_2$	Repeated index indicates summation
$\prod_{i=1}^{3} a_i = a_1 a_2 a_3$	Product indexing
$\mathbf{A} = \begin{bmatrix} a_1 \\ a_2 \\ a_3 \end{bmatrix}$	Index identifies matrix elements
$\mathbf{A} = \begin{bmatrix} a_{11} & a_{12} & a_{13} \\ a_{21} & a_{22} & a_{23} \\ a_{31} & a_{32} & a_{33} \end{bmatrix}$	Double index identifies matrix elements

Curves and Surfaces

Here we see examples of vectors, matrices, and indexing applied to define curves and surfaces. Vector and matrix notation for Hermite, Bézier, and B-spline curves and surfaces is far from uniform and standardized in the literature. The notation in this text conforms to generally accepted, although not universal, forms. Also note that not all these forms appear in this textbook. And there are many not included here.

$\mathbf{p}(u)=\sum_{i=0}^{n}\mathbf{b}_i F_{i,n}(u)$	Hermite cubic curve, $n = 3$
$\mathbf{P}(u)=\mathbf{U}\mathbf{M}_{\mathrm{F}}\mathbf{B}$	Hermite cubic curve, $n = 3$
$\mathbf{p}(u)=\sum_{i=0}^{n}\mathbf{p}_i B_{i,n}(u)$	Bézier cubic curve, $n = 3$
$\mathbf{P}(u)=\mathbf{U}\mathbf{M}_{\mathrm{B}}\mathbf{P}$	Bézier curve in matrix notation
$\mathbf{p}(u)=\sum_{i=0}^{n}\mathbf{p}_i N_{i,K}(u)$	B-spline curve
$\mathbf{P}_i(u)=\mathbf{U}\mathbf{M}_N\mathbf{P}_K$	B-spline curve (matrix notation)
$\mathbf{p}(u)=\dfrac{\sum_{i=0}^{n}h_i\mathbf{p}_i N_{i,K}(u)}{\sum_{i=0}^{n}h_i N_{i,K}(u)}$	Rational B-spline curve
$\mathbf{p}(u,w)=\sum_{i=0}^{m}\sum_{j=0}^{n}\mathbf{b}_{ij}F_{i,m}(u)F_{j,n}(w)$	Hermite surface
$\mathbf{P}(u,w)=\mathbf{U}\mathbf{M}_F\mathbf{B}\mathbf{M}_F^{\mathrm{T}}\mathbf{W}^{\mathrm{T}}$	Hermite surface: matrix notation
$\mathbf{p}(u,w)=\sum_{i=0}^{m}\sum_{j=0}^{n}\mathbf{p}_{ij}B_{i,m}(u)B_{j,n}(w)$	Bézier surface
$\mathbf{P}(u,w)=\mathbf{U}\mathbf{M}_B\mathbf{P}\mathbf{M}_B^{\mathrm{T}}\mathbf{W}^{\mathrm{T}}$	Bézier surface: matrix notation
$\mathbf{p}(u,w)=\sum_{i=0}^{m}\sum_{j=0}^{n}\mathbf{p}_{ij}N_{i,K}(u)N_{j,L}(w)$	B-spline surface
$\mathbf{P}_{st}(u,w)=\mathbf{U}\mathbf{M}_S\mathbf{P}_{KL}\mathbf{M}_S^{\mathrm{T}}\mathbf{W}^{\mathrm{T}}$	B-spline surface: matrix notation
$\mathbf{p}(u,w)=\dfrac{\sum_{i=0}^{m}\sum_{j=0}^{n}h_{ij}\mathbf{p}_{ij}N_{i,K}(u)N_{j,L}(w)}{\sum_{i=0}^{m}\sum_{j=0}^{n}h_{ij}N_{i,K}(u)N_{j,L}(w)}$	Rational B-spline surface

2 The Visual Vector

This chapter is about how to visualize a vector and to see it as a geometric object. Scientists, engineers, and mathematicians see and work with vectors almost as if they were real objects. In fact, vectors have characteristics that are constant when they are moved about, at least in a mathematical sense. More often than not, it is useful to visualize a vector as a directed line segment, and this is where we start.

2.1 Directed Line Segment

Sometime during the 19th century, mathematicians developed a new mathematical object—a new kind of number. They were motivated in part by an important observation: Physicists had long known that while some phenomena can be described by a single number—the temperature of a beaker of water (6°C), the mass of a sample of iron (17.5 g), or the length of a rod (31.736 cm)—other phenomena require something more:

A ball strikes the rail cushion of a billiard table at a certain speed and angle (Figure 2.1); we cannot describe its initial trajectory or its rebound by a single number.

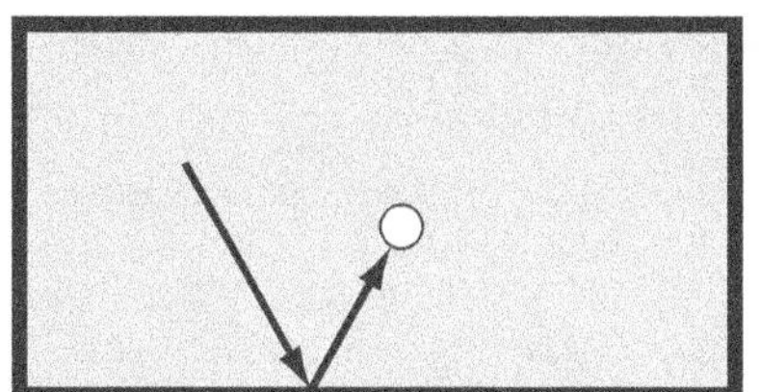

Figure 2.1 Rebound of a billiard ball

A pilot flies a plane due west with an air speed of 800 kph in a north crosswind of 120 kph (Figure 2.2); we cannot describe the plane's true motion relative to the ground by a single number. We would like to know the plane's speed and the true heading of its path.

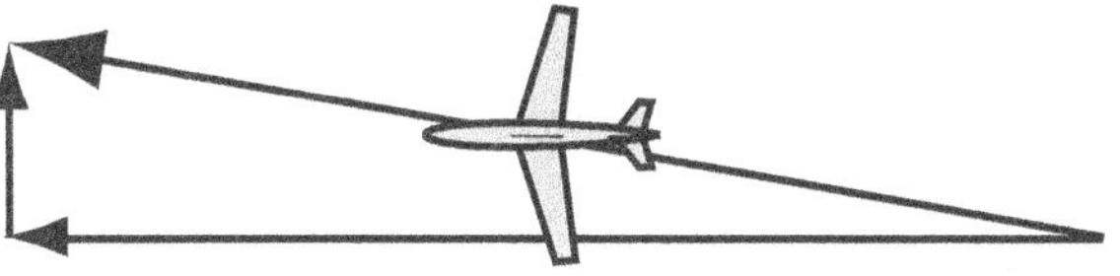

Figure 2.2 Flight path of a plane in a crosswind

Two elastic bodies collide; if we know the momentum (mass × velocity) of each body before impact, then we can determine their speed and direction after impact (Figure 2.3). The description of the momentum of each body seems to require more information than is contained in a single number.

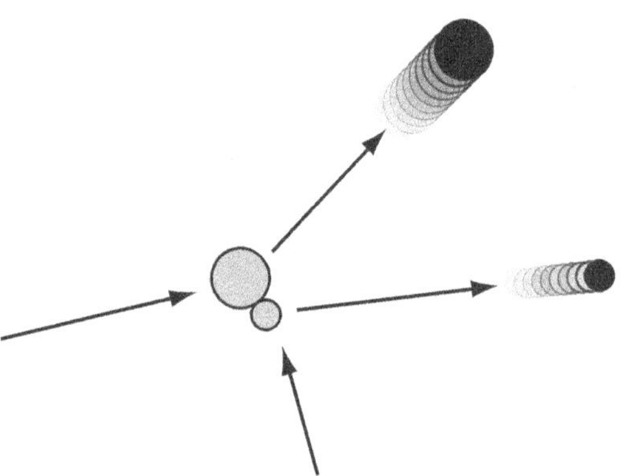

Figure 2.3 Collision

Billiard balls, airplanes, and colliding bodies need a mathematical object that describes both the speed and the direction of their motion. Mathematicians found that they could do this by using an ordered set of numbers made up of two or more numbers, or components. When the components are combined according to certain rules, the results produce both magnitude and direction.

For the example of the airplane flying in a crosswind, its magnitude is given in kilometers per hour, and its direction is given by a compass reading. For any specified elapsed time t and velocity v, the problem immediately reduces to one of directed distance or displacement d, because $d = vt$ (distance = velocity × time). When a set of numbers is associated with a distance and direction in these examples, it is a vector, and we use vectors to solve these kinds of problems. And much more . . . components can be more than just numbers. They can be mathematical functions . . . as we shall see in the chapters to follow.

So what does a vector look like? First, the great and useful thing about vectors is that we can visualize them by drawing pictures of them as arrows or directed line segments. Figure 2.4 shows several vectors with different lengths and directions. A boldface lowercase letter identifies each one.

An arrow has the properties of length and direction, and this is a good way to think of a vector. The length of the arrow represents the magnitude of a vector, and the arrow's orientation represents the direction of a vector. A vector may have the same magnitude and direction as that of some other vector or a different magnitude and direction from that of the vector. Figure 2.4 shows several examples lying in the plane of the paper. Two vectors are equal if they have the same length and direction. If we can move vector **a** so that it remains parallel to its original orientation into coincidence with another vector **b**, then **a** and **b** are equal. Although **c** is the same length as **a**, and parallel to it, it is in the opposite direction, so they are not equal. We also see that **d** and **e** are not the equivalent of **a**. The vectors **e** and **f** are in the same direction but have different lengths, so they are not equal.

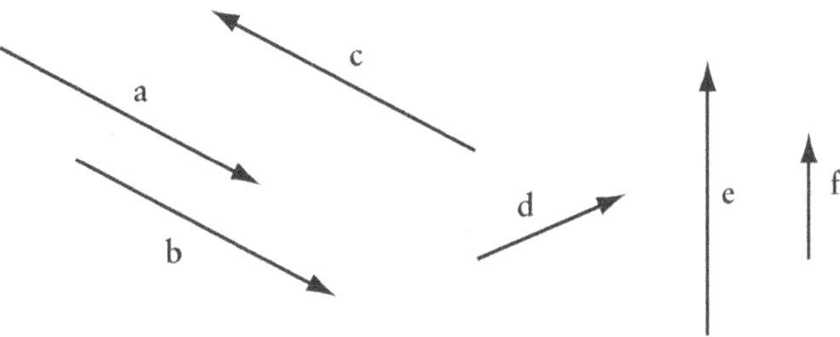

Figure 2.4 Vectors as directed line segments

Now let's imagine that an ordered pair of numbers is a set of instructions for moving about on a flat two-dimensional surface. For the moment, let's agree that the first number represents a displacement (how we are to move) east if plus (+) or west if minus (–), and that the second number represents a displacement north (+) or south (–). For example, we can interpret the set of numbers (16.3, –10.2) as a displacement of 16.3 units of length (feet, meters, light-years, or whatever) to the east, followed by a displacement of 10.2 units of length (same as the east-west units) to the south (Figure 2.5).

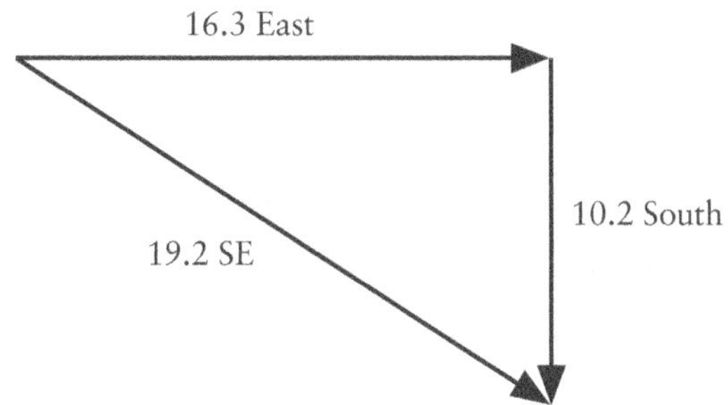

Figure 2.5 Displacements in the plane

Notice something interesting here. By applying the Pythagorean theorem, we find that this is equivalent to a displacement of $\sqrt{(16.3)^2 + (-10.2)^2} = 19.23$ units of length in a southeasterly direction. We can be more precise about the direction. The compass heading from our initial position would be

$$90° + \arctan(10.2/16.3) = 122°\text{SE}$$

We have added the two displacements to produce the resultant displacement. This is key to understanding how to add vectors to produce a single equivalent vector.

We can apply this interpretation to any ordered pair of numbers. Each pair produces both a magnitude (the total displacement) and a direction. "Ordered" in this case means that we assign a unique and consistent direction to each displacement, depending on whether it is the first or second of the ordered pair.

We treat ordered triples of numbers (a_1, a_2, a_3) in the same way, by creating a third dimension to complement the compass headings of the two-dimensional plane. This is easy enough: Adding up (+) and down (–) is all that we need to do. We have, then, for an ordered triple, an agreement that the first number represents a displacement east (+) or west (–), the second number a displacement north (+) or south (–), and the third number a displacement up (+) or down (–) (Figure 2.6). Later we will see how the three numbers of an ordered triple also can be made to yield a resultant distance and direction. And we will also assign a more general meaning to these three ordered numbers: They take on the meaning of Cartesian coordinates x, y, and z.

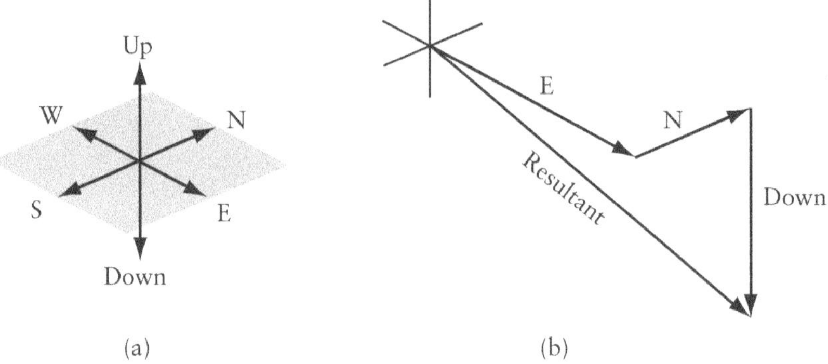

Figure 2.6 Displacement in three dimensions

Here again we add vectors, this time three, to produce the resultant vector.

2.2 Free and Fixed Vectors

If a vector is not assigned or constrained to a particular location, then it is a free vector. We can move a free vector around, parallel to its initial position, without changing its length and orientation. This is true in both two and three dimensions. We see that it is the same vector no matter where we place it.

Vectors representing problems in pure displacements need only to use unrestricted parallel transport to find a solution that is itself given in terms of free vectors. However, many problems require what we call a fixed vector, which is often the case in physics, computer animation, and geometric and 3D modeling.

Fixed vectors begin at a common point, usually the origin of a coordinate system. For the familiar two-dimensional Cartesian coordinate system, the components of a fixed vector, whose tail is at the origin, lie along the principal axes, and their length or magnitudes correspond to the coordinates of the point at its arrowhead tip (Figure 2.7).

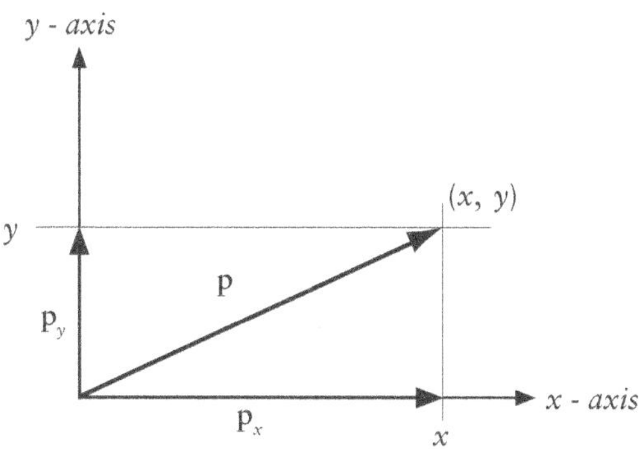

Figure 2.7 Fixed vector in two dimensions

The fixed vector **p** has component vectors $\mathbf{p}_x$ and $\mathbf{p}_y$, usually corresponding to the x and y coordinates. This means that $\mathbf{p} = \mathbf{p}_x + \mathbf{p}_y$, and in three dimensions $\mathbf{p} = \mathbf{p}_x + \mathbf{p}_y + \mathbf{p}_z$ (Figure 2.8). More on this in the next section.

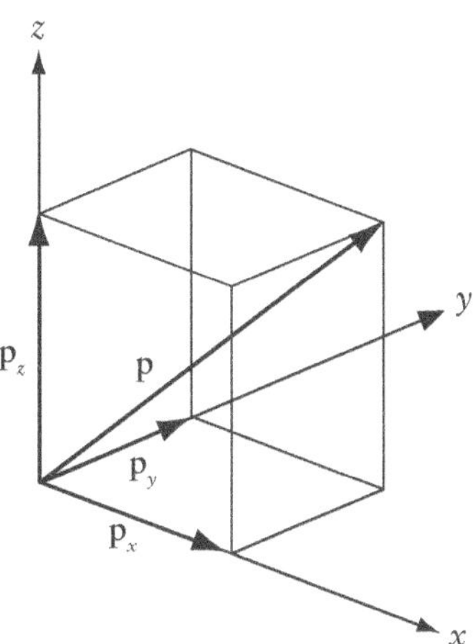

Figure 2.8 Fixed vector in three dimensions

The nature of a particular problem may blur the distinction between free and fixed vectors. Also, most arithmetic and algebraic operations are identical for both kinds of vectors. So the distinction may be as important for visualization and intuition as for any other reason.

2.3 Vector Addition

We use the idea of parallel transport of directed line segments to add two vectors **a** and **b** as follows: Transport **b** until its tail is coincident with the head of

a (Figure 2.9). Their sum, **a** + **b**, is the directed line segment beginning at the tail of **a** and ending at the head of **b**. Transporting **a** so that its tail coincides with the head of **b** produces the same result: **a** + **b** = **b** + **a**. So we see that the order in which we add two vectors is not important. Later we'll discuss how to add two vectors by adding their components. For now, we'll look at the geometric interpretation of addition.

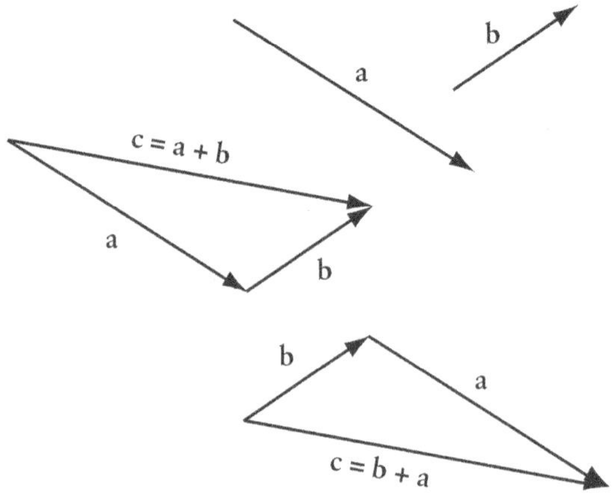

Figure 2.9 Vector addition using parallel transport

We extend this method to adding many vectors by simply transporting each vector to form a head-to-tail chain (Figure 2.10). Connecting the tail of the first vector to the head of the last vector in the chain produces the resultant vector. In the figure, this construction yields **e** = **a** + **b** + **c** + **d**. This works in both two and three dimensions.

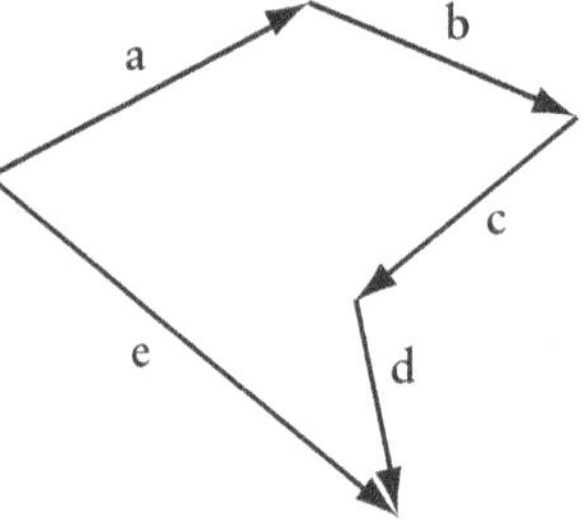

Figure 2.10 Vector addition using a head-to-tail chain of vectors

The idea of parallel transport of free vectors leads directly to the parallelogram law for adding two vectors, as well as for finding a vector's components. First, transport the vectors **a** and **b** so that their tails coincide, and then complete the construction of the suggested parallelogram $ABCD$ (Figure 2.11). The diagonal **c**, or AC, represents their sum; that is, **c** = **a** + **b**. More importantly, as we will see next, it is possible to think of the vectors **a** and **b** as components of **c**.

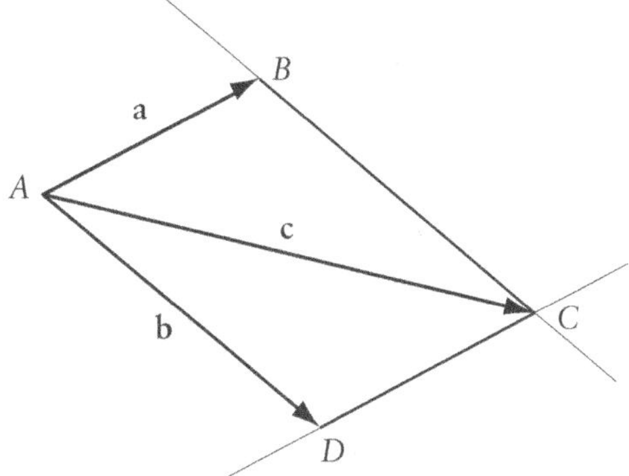

Figure 2.11
The parallelogram law

Notice that we need only half of the parallelogram, for example, *ABC* (Figure 2.11). Using parallel transport, move the vector **b** so that its tail joins the head of vector **a**. As a result, the vector **b** lies along the line *BC*. This is the classic construction for the addition of two vectors, in this case **a** + **b**.

2.4 Vector Components

So what are the components of a vector? Let's start with vectors in two dimensions. Any vector **a** can be broken down into two components, **b** and **c**, if and only if they add up to the original vector, meaning **a** = **b** + **c**. These three vectors must not be collinear but otherwise may have any orientation and point in any direction. Most commonly, the components **b** and **c** will emanate from the origin of a coordinate system and lie along its axes. When we study basis vectors in Chapter 6, we will see that the coordinate system is not necessarily the familiar orthogonal Cartesian one.

And what works in two dimensions also works in three or more dimensions. In the case of three-dimensional space, a vector breaks down into three component vectors, which must be noncollinear. Thus, we might have **a** = **b** + **c** + **d**. It follows that any vector can be one of the components of any other vector. For any vector in two dimensions, we are free to pick any other vector as one of its components as long as the remaining component satisfies the rule of addition offered above. For example, if for a vector **a** we choose **b**, then the remaining component is determined by **c** = **a** – **b**. In three dimensions, we may select two components, with the third one determined by **d** = **a** – **b** – **c**.

A vector in two-dimensional space has two and only two components. Although the two components are not unique, they must always add up to produce the original vector. Similarly, this is true for a three-dimensional vector, which has three and only three components. A further restriction prohibits components from being parallel. Another way to think of this is that the components of a vector each must point in a different direction. (Section 3.7 discusses linear vector spaces and explains this restriction.)

The parallelogram law of addition suggests a way to find the components of a vector **a** along any two directions L and M (Figure 2.12). We construct lines L and M through the tail of **a** and their parallel images L' and M' through the head of **a**. This construction produces a parallelogram whose adjacent sides $\mathbf{a}_L$ and $\mathbf{a}_M$ are the components of **a** along L and M, respectively. We see, of course, that these components are are not unique! We could just as easily construct other lines, say P, Q and P', Q', to find $\mathbf{a}_P$ and $\mathbf{a}_Q$. Obviously, $\mathbf{a} = \mathbf{a}_P + \mathbf{a}_Q$, and, in general, $\mathbf{a}_P \neq \mathbf{a}_L$, $\mathbf{a}_M$ or $\mathbf{a}_Q \neq \mathbf{a}_L$, $\mathbf{a}_M$. This property of the non-uniqueness of the vector's components is a very powerful feature, which we will see in the chapters to follow.

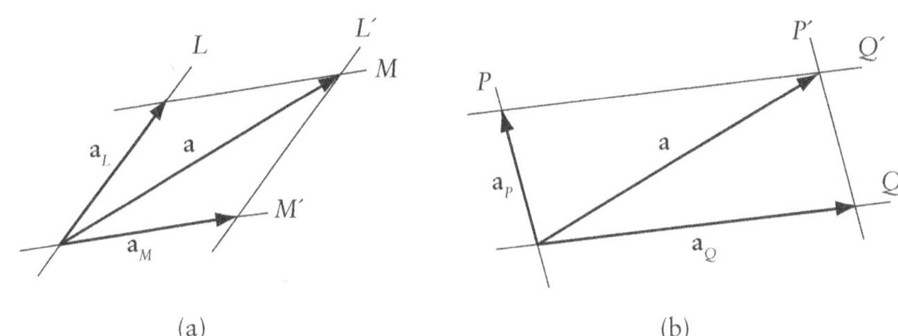

Figure 2.12 Vector components in two arbitrary directions

Again, because we have not constrained these vectors to any particular location, they are free vectors. We have moved them around and preserved their properties of length and orientation. This is possible only if we always move them parallel to themselves. It is also true in three dimensions (Figure 2.13). Notice that the components are not and must not be coplanar. In the figure, we use them to construct a rectangular parallelepiped.

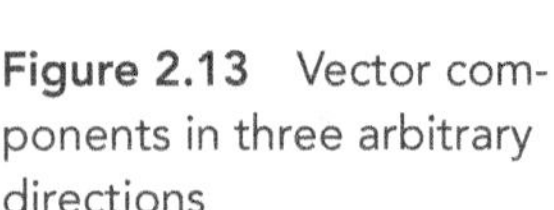

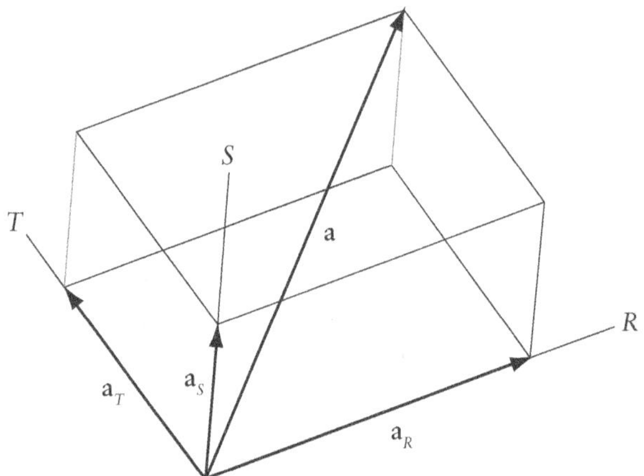

Figure 2.13 Vector components in three arbitrary directions

2.5 Vector Multiplication

There are two very different ways to multiply two vectors, **a** and **b**. (We postpone the discussion of the product of three vectors until Chapter 3.) One is the scalar product (also known as the dot product or inner product).

And the other is the vector product. Here we look at the geometric interpretation of these products, albeit somewhat handicapped by doing this without their algebraic definition.

We write the scalar product like this: **a** • **b**. Later we will see how to express this in terms of the components and what meaning we can get from its algebraic definition. If the angle between these two vectors is θ, then we define the scalar product as follows:

$$\mathbf{a} \bullet \mathbf{b} = (\text{length of } \mathbf{a}) \times (\text{length of } \mathbf{b}) \cos\theta$$

or

$$\mathbf{a} \bullet \mathbf{b} = |\,\mathbf{a}\,||\,\mathbf{b}\,| \cos\theta. \tag{2.1}$$

The result is a scalar, where $|\,\mathbf{a}\,|$ and $|\,\mathbf{b}\,|$ are the lengths of **a** and **b**, respectively. They are always positive. At this point, it seems that this definition just pops up out of thin air. Later we will see how it arises within the algebraic definition, when certain rules are followed (Chapter 3). Remember that a scalar is just a number without the properties of a vector, although it could have some physical characteristic associated with it, such as length, temperature, and so forth.

We obviously can use Equation 2.1 to find the angle between two vectors **a** and **b**:

$$\theta = \cos^{-1}\left(\frac{\mathbf{a} \bullet \mathbf{b}}{|\,\mathbf{a}\,||\,\mathbf{b}\,|}\right). \tag{2.2}$$

This means that the scalar projection of **a** onto **b** (Figure 2.14) is $|\,\mathbf{a}\,| \cos\theta$. And also the vector projection of **a** onto **b** is $|\,\mathbf{a}\,|\,\hat{\mathbf{b}} \cos\theta$, where $\hat{\mathbf{b}}$ is a unit vector in the direction of **b**. A unit vector's length or magnitude is 1 (more on unit vectors in Chapter 3).

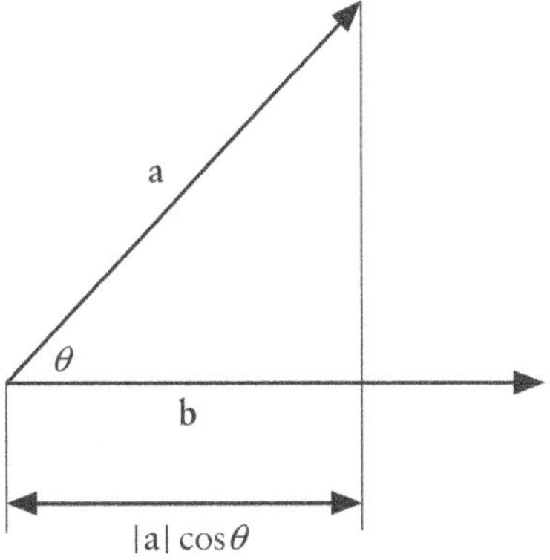

Figure 2.14 Scalar and vector projection of a onto b

We see that **a** and **b** are perpendicular if

$$\mathbf{a} \bullet \mathbf{b} = 0. \tag{2.3}$$

If they are parallel, then

$$\mathbf{a} \bullet \mathbf{b} = |\,\mathbf{a}\,||\,\mathbf{b}\,|. \tag{2.4}$$

Here is another way to multiply two vectors: the vector product. We write the vector product (also known as the cross product) of **a** and **b** like this: **a** × **b**. This produces another vector, say **c**, that turns out to be perpendicular to the plane containing **a** and **b**. (Be patient. Section 3.5 shows why this is true.) So you will see equations like this:

$$\mathbf{a} \times \mathbf{b} = \mathbf{c}. \tag{2.5}$$

Figure 2.15 shows a way to visualize this.

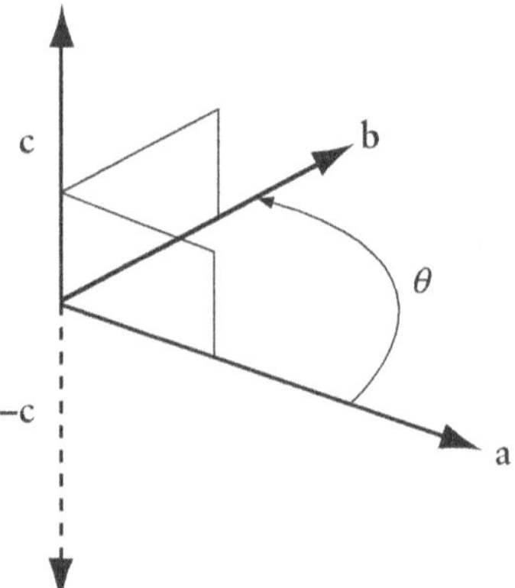

Figure 2.15 Vector product of two vectors

The order in which the vector product is taken determines the direction of **c**, so that

$$\mathbf{b} \times \mathbf{a} = -\mathbf{c}. \tag{2.6}$$

When this product is developed in terms of the components, it will be clear that the vector produced by **a** × **b** is in the opposite direction of **b** × **a**. Think about rotating **a** into **b** through the angle between them, θ, while curling the fingers of your right hand through that rotation. Your thumb now points in the direction of the resultant vector, **c**. Mathematicians call this the right-hand rule. If instead you cross **b** into **a**, curling the fingers of your right hand in that sense, then the thumb points in the direction opposite **c**; in other words, it points as does –**c**, shown in the figure.

The vector product has another interesting geometric characteristic (Figure 2.16). The magnitude of the vector product is equal to the area of the parallelogram formed by the two vectors, so that

$$A_{\square} = |\mathbf{a} \times \mathbf{b}| = |\mathbf{b} \times \mathbf{a}|. \tag{2.7}$$

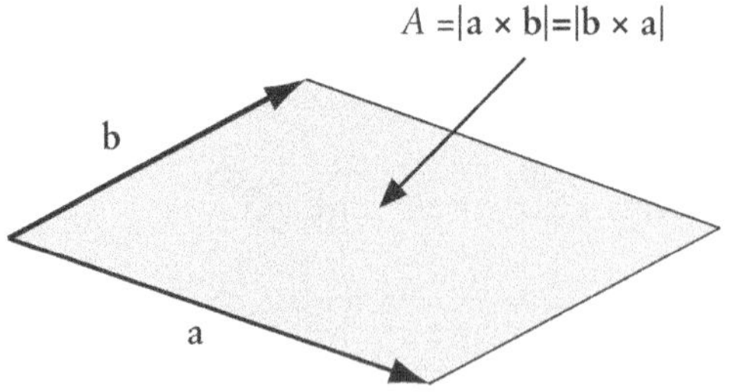

Figure 2.16 Area of a parallelogram formed by the vector product

After studying Chapter 3, come back here and prove this.

2.6 A Geometry Problem

Given four noncoplanar points in space forming the vertices of a skew polygon (Figure 2.17), use vectors to show that the midpoints of the edges forming it are coplanar and are the vertices of a parallelogram.

First, denote the polygon's edges as the vectors **a**, **b**, **c**, and **d**, and denote the midpoints of the edges as A, B, C, and D. We want to demonstrate that the vectors $\overline{AB}$ and $\overline{CD}$ are parallel. If this is true, then we conclude that they are coplanar and that points A, B, C, and D all lie in that plane. Furthermore, we see that $|\overline{AB}|=|\overline{CD}|$ and are opposite sides of a parallelogram. (Notice that we have introduced another way to denote a vector—with an overline, or bar over the top: $\overline{AB}$ and $\overline{CD}$, for example.)

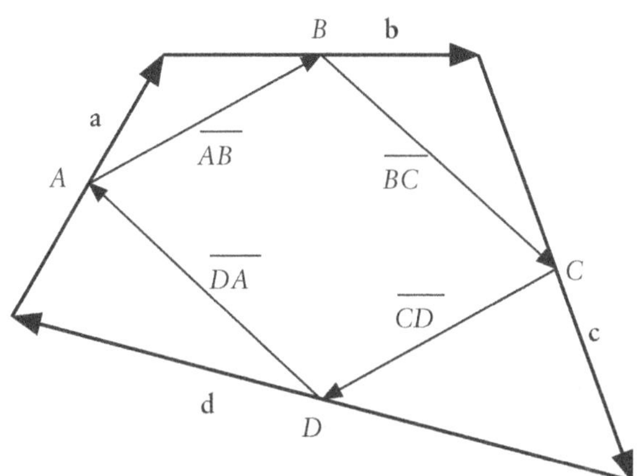

Figure 2.17 A geometry problem

From the figure we see that

$$\mathbf{a}+\mathbf{b}+\mathbf{c}+\mathbf{d}=0, \tag{2.8}$$

$$\overline{AD}+(\mathbf{a}/2)+(\mathbf{d}/2)=0, \tag{2.9}$$

and

$$\overline{BC}-(\mathbf{c}/2)-(\mathbf{d}/2)=0. \tag{2.10}$$

Rearranging the last two equations, we find

$$\mathbf{a}=-2\overline{AD}-\mathbf{d} \tag{2.11}$$

and

$$\mathbf{b}+\mathbf{c}=2\overline{BC}. \tag{2.12}$$

Substituting these results into $\mathbf{a}+\mathbf{b}+\mathbf{c}+\mathbf{d}=0$, we find

$$-2\overline{AD}-\mathbf{d}+2\overline{BC}+\mathbf{d}=0 \tag{2.13}$$

and

$$\overline{AD}=\overline{BC}. \tag{2.14}$$

Thus $\overline{AD}$ and $\overline{BC}$ are parallel and coplanar, and A, B, C, and D lie in a common plane. Finally, we see it must be true that $\overline{AB} = \overline{DC}$ and that they form a parallelogram with $\overline{AD}$ and $\overline{BC}$.

2.7 The Idea of a Vector as a Displacement

This chapter introduced the idea of a vector as a directed line segment, or more simply as a displacement. Vectors are not motions or motion paths. We can think of a vector as describing the relationship between point A and point B. The vector tells us that point B is a certain distance and direction from point A (Figure 2.18).

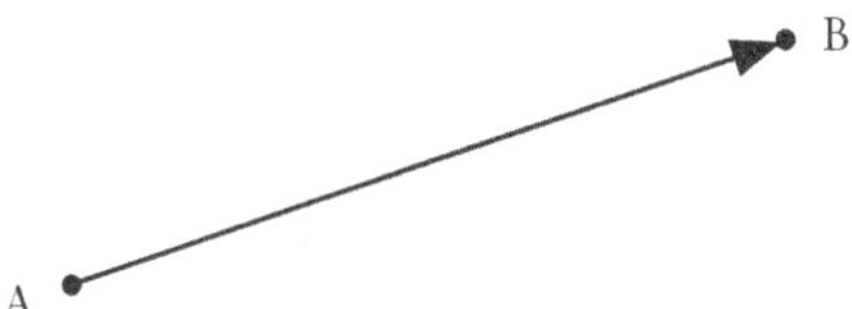

Figure 2.18 A vector as a displacement from A to B

Later we will see that given the coordinates of points A and B, we can find the components of the vector connecting them. In fact, we will also see that vector functions (mathematical equations) may be used to define not just individual points, but also the locus of points that define curves and surfaces . . . but more on this later. So the idea here is that vectors often look like displacements, and that this outlook helps us to understand their usefulness to geometric and 3D modeling.

3 The Algebraic Vector

This chapter builds on the visual interpretation of a vector discussed in Chapter 2. It is about the mathematically more powerful algebraic description of a vector and vector operations. Here is where you will begin to see the value of vectors in geometric and 3D modeling: how a fixed vector represents a point whose coordinates are the vector's components, how the products of vectors produce useful geometric information.

3.1 There's a Vector in Your Coordinate System

Yes, there are vectors deep within your coordinate system ... in your familiar rectangular coordinate system. (And in many other coordinate systems, too, and these are discussed later.) Let's start with the familiar two-dimensional Cartesian coordinate system, placing a vector **a** along the x axis, with its tail at the origin, and a vector **b** along the y axis, with its tail also at the origin (Figure 3.1). We find their sum, **c** = **a** + **b**, by completing the rectangular parallelogram.

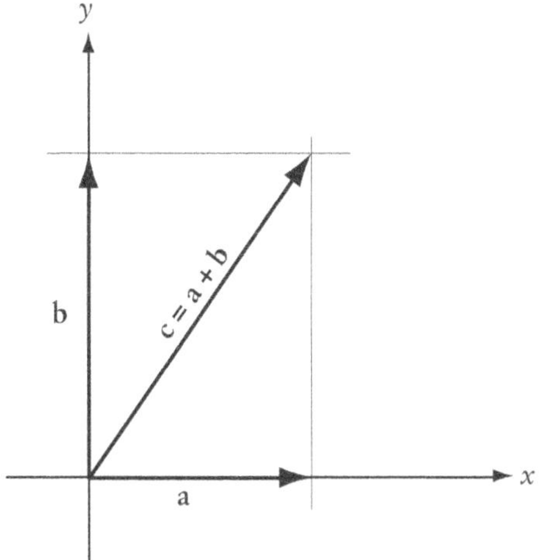

Figure 3.1 Defining a coordinate system with vectors

Now let's think of **a** and **b** as components of **c**. We see that the magnitudes of components **a** and **b** of **c** (written as | **a** | or a and | **b** | or b, respectively) are also the coordinates of the arrowhead point, x_c, y_c (Figure 3.2). The subscript c tells us that x_c and y_c refer to **c**.

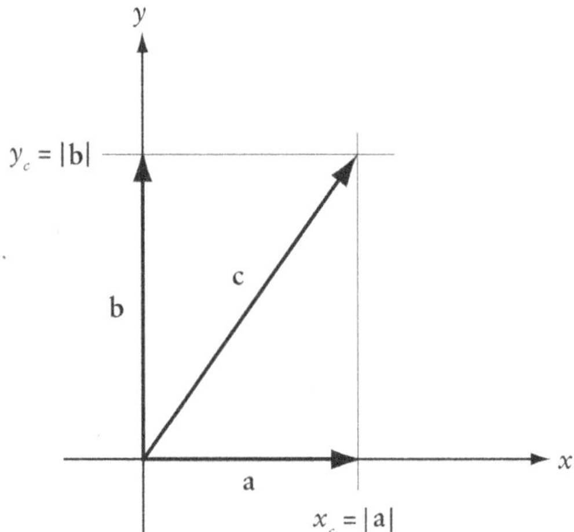

Figure 3.2 Components to coordinates

Let's reverse this process. Given a vector **c**, with its tail at the origin, through the point of its arrowhead, construct lines *L* and *M* parallel to the *x* and *y* axes, respectively. Use the resulting points of intersection, *A* and *B*, with the *x* and *y* axes to lay out the vector's components **a** and **b** (Figure 3.3). And so we see again that the sum of a vector's components is equal to the vector itself. And the coordinates of the point at the arrowhead are *a*, *b*. This also applies to vectors in three-dimensional coordinate systems.

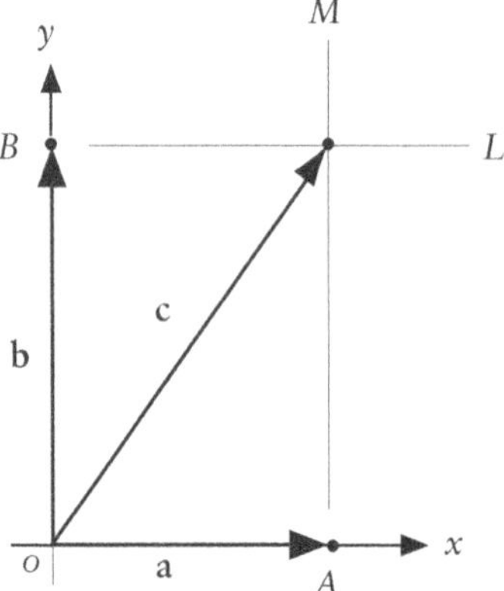

Figure 3.3 Coordinates to components

What about the magnitudes of **a**, **b**, and **c**? Remember, magnitude is equal to the length of the directed line segment. It is a real number and always positive, as lengths always are (with the exception of the null vector, whose length is zero). The lengths of **a** and **b** are *OA* and *OB*, respectively.

Here are some ways to write the magnitude of any vector **a**: $\|\mathbf{a}\|$, $|\mathbf{a}|$, or simply *a*. The Pythagorean theorem gives the magnitude of **c**:

$$\|\mathbf{c}\| = \sqrt{\|a\|^2 + \|b\|^2} \text{ or } |\mathbf{c}| = \sqrt{|\mathbf{a}|^2 + |\mathbf{b}|^2} \text{ or } c = \sqrt{a^2 + b^2}. \tag{3.1}$$

If

$$|\mathbf{c}| = 1, \tag{3.2}$$

then **c** is a unit vector (see Section 3.3 on unit vectors).

3.2 Form and Equality

Here is one of the forms we will use to represent a vector:

$$\mathbf{p} = p_x\mathbf{i} + p_y\mathbf{j} + p_z\mathbf{k}, \tag{3.3}$$

where **i**, **j**, and **k** are mutually orthogonal vectors of unit length that mathematicians call basis vectors (Figure 3.4). They establish the basis of a coordinate system. In general, basis vectors need not be mutually orthogonal or of unit magnitude (for more on basis vectors, see Chapter 6).

Now seems a good time to bring up two terms that are often used incorrectly: orthogonal and orthonormal. Their definitions are:

- *Orthogonal basis vectors* are mutually perpendicular to one another. There is no restriction on their lengths. That is, the length of each basis vector may be different from the others.
- *Orthonormal basis vectors* are not only mutually perpendicular to one another; each one is also a unit vector.

The terms p_x, p_y, and p_z are displacements in the x, y, and z directions, respectively. They are the magnitudes of the three vectors $p_x\mathbf{i}$, $p_y\mathbf{j}$, and $p_z\mathbf{k}$ that, added together, produce **p**.

Usually it is just simpler to write

$$\mathbf{p} = (p_x, p_y, p_z), \tag{3.4}$$

omitting **i**, **j**, and **k** and their addition, which are implied.

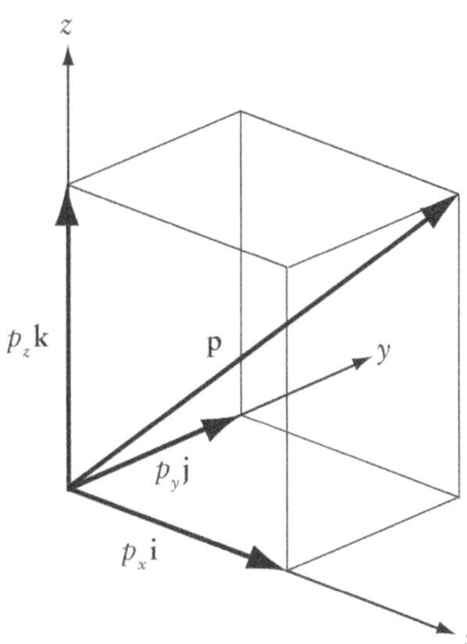

Figure 3.4 Visualizing Equation 3.4

Two vectors **a** and **b** are equal if and only if their respective components are equal. Thus, if $a_x = b_x$, $a_y = b_y$, and $a_z = b_z$, then **a** = **b**. This means that equal vectors have equal magnitudes, equivalent directions, and, if they are fixed, a common starting point. Free vectors are equal if they can be made to coincide by a translation.

In matrix form (more about this in Chapter 4), Equation 3.3 becomes

$$\mathrm{P} = \begin{bmatrix} p_x & p_y & p_z \end{bmatrix} \begin{bmatrix} \mathbf{i} \\ \mathbf{j} \\ \mathbf{k} \end{bmatrix}, \tag{3.5}$$

or, alternatively,

$$\mathrm{P} = \begin{bmatrix} \mathbf{i} & \mathbf{j} & \mathbf{k} \end{bmatrix} \begin{bmatrix} p_x \\ p_y \\ p_z \end{bmatrix}. \tag{3.6}$$

To simplify the matrix form, we may omit the **i**, **j**, and **k**, so that

$$\mathrm{P} = \begin{bmatrix} x & y & z \end{bmatrix} \text{ or } \begin{bmatrix} x \\ y \\ z \end{bmatrix}. \tag{3.7}$$

This is particularly appropriate for a fixed vector, where we interpret the components, now presented by matrix elements, as point coordinates.

In the tensor form (for those readers who are interested), the superscript on x^i in Equation 3.8 below identifies the components as $x^1, x^2, \ldots, x^n$, where the $\mathbf{e}_i$ are the corresponding basis vectors. In general, the components do not correspond to those of an orthonormal Cartesian space (Chapter 6 is all about basis vectors). For a three-dimensional space,

$$\begin{aligned} \mathbf{x} &= \sum_{i=1}^{3} x^i \mathbf{e}_i \\ &= x^1\mathbf{e}_1 + x^2\mathbf{e}_2 + x^3\mathbf{e}_3, \end{aligned} \tag{3.8}$$

or

$$\mathbf{x} = x^i \mathbf{e}_i. \tag{3.9}$$

The repetition of the index i indicates summation. This is the Einstein convention used in tensor analysis. It is a handy shorthand that often shows up in curve or surface formulations, and it is the simplest expression of a Cartesian tensor (not discussed here). Chapter 6 makes good use of this notation toward developing the idea of basis vectors.

3.3 Magnitude and Direction

The last chapter presented a visual representation of a vector as an arrow-headed line segment, with a definite length and pointing in a definite direction. Here we will see how to extract these characteristics from a mathematical definition of a vector, using slightly different notation than that used in previous sections.

Magnitude

The magnitude of the vector $\mathbf{p} = (p_x, p_y, p_z)$ is the real number symbolized by $|\mathbf{p}|$, where

$$|\mathbf{p}| = \sqrt{p_x^2 + p_y^2 + p_z^2}\,. \tag{3.10}$$

Again, this is simply a statement of Pythagoras's theorem in an ordinary space of three dimensions (Figure 3.5).

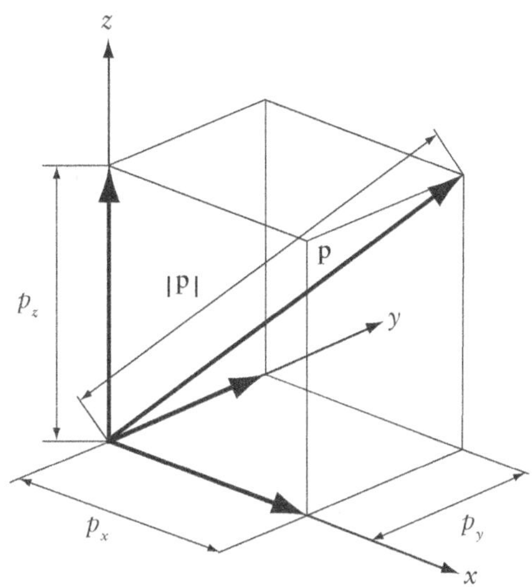

Figure 3.5 Vector magnitude and Pythagoras's theorem

It is always the case that $|\mathbf{p}| \geq 0$. Furthermore, $|\mathbf{p}| = 0$ if and only if $\mathbf{p} = 0$; that is, $\mathbf{p} = (0, 0, 0)$, which is the null vector.

If $\mathbf{a} = (2, 3, 5)$ and $\mathbf{b} = (6, -1, 3)$, then

$$|\mathbf{a}| = \sqrt{2^2 + 3^2 + 5^2} = 6.16$$

and

$$|\mathbf{b}| = \sqrt{6^2 + (-1)^2 + 3^2} = 6.78.$$

Multiplying a vector by a scalar k changes its magnitude; thus,

$$k\mathbf{p} = (kp_x, kp_y, kp_z) \tag{3.11}$$

and

$$\begin{aligned} |k\mathbf{p}| &= k\sqrt{p_x^2 + p_y^2 + p_z^2} \\ |k\mathbf{p}| &= k\,|\mathbf{p}|. \end{aligned} \tag{3.12}$$

Notice that if $k < 0$, the direction of $\mathbf{p}$ is reversed. Later we will see that the magnitude of a vector is invariant under rigid-body transformations; that is, magnitude is independent of direction. Here is a summary of the possible effects of a scalar multiplier k:

- $k > 1$ increases length and preserves direction.
- $k = 1$ means no change in length or direction.
- $0 < k < 1$ decreases length and preserves direction.
- $k = 0$ produces the null vector with direction undefined.
- $-1 < k$ decreases length and reverses direction.
- $k = -1$ means no change in length and reverses direction.
- $k < -1$ reverses direction and increases length.

The concept of vector magnitude has a stealthy complication. We must account for what the magnitude measures: for example, a dimensionless real number, distance, force, velocity, acceleration, and more. For more about this, jump to Section 3.4 on vector addition and subtraction.

Direction

The direction cosines (or direction numbers) of a vector determine its direction. For the vector **p** (from Figure 3.6), the direction cosines are $\cos\alpha,\ \cos\beta,\ \cos\gamma$, where

$$\cos\alpha = \frac{p_x}{|\mathbf{p}|}, \quad \cos\beta = \frac{p_y}{|\mathbf{p}|}, \quad \cos\gamma = \frac{p_z}{|\mathbf{p}|}. \tag{3.13}$$

Since $|\mathbf{p}| = \sqrt{p_x^2 + p_y^2 + p_z^2}$, then $\left(\frac{p_x}{|\mathbf{p}|}\right)^2 + \left(\frac{p_y}{|\mathbf{p}|}\right)^2 + \left(\frac{p_z}{|\mathbf{p}|}\right)^2 = 1$, so that

$$\cos^2\alpha + \cos^2\beta + \cos^2\gamma = 1. \tag{3.14}$$

Equation 3.14 tells us that any two direction cosines are sufficient to determine the direction of a vector. Certain forms of Hermite cubic curves use this to add an extra degree of freedom in controlling the shape of a curve via its tangent vectors.

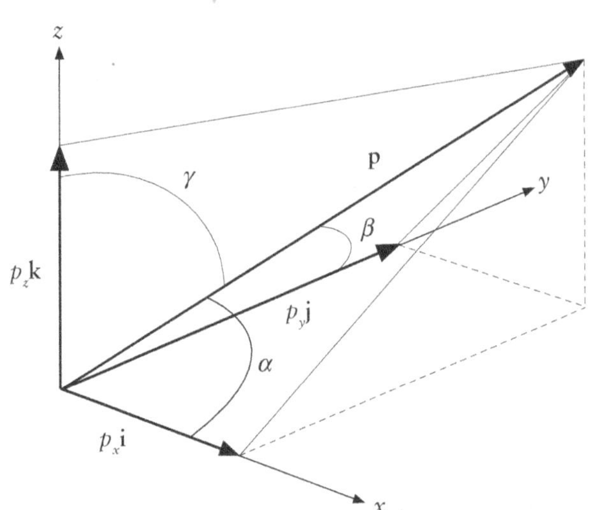

Figure 3.6 Direction angles and vectors used for direction cosine computation

Verify the direction cosines of the following vectors (the notation **â** indicates the unit vector of **a**, whose components are its direction cosines):

a. $\mathbf{a} = [\,5 \quad 6\,]$ $\quad\hat{\mathbf{a}} = [\,0.640 \quad 0.768\,]$

b. $\mathbf{b} = [\,5 \quad -5\,]$ $\quad\hat{\mathbf{b}} = [\,0.707 \quad -0.707\,]$

c. $\mathbf{c} = [\,0 \quad 7\,]$ $\quad\hat{\mathbf{c}} = [\,0 \quad 1\,]$

d. $\mathbf{d} = [\,-7 \quad 0\,]$ $\quad\hat{\mathbf{d}} = [\,-1 \quad 0\,]$

e. $\mathbf{e} = [\,-5 \quad 3\,]$ $\quad\hat{\mathbf{e}} = [\,0.857 \quad 0.514\,]$

Notice that if α, β, and γ are the angles between **a** and the x, y, and z axes, respectively, then

$$\hat{a}_x = \frac{a_x}{|\mathbf{a}|} = \cos\alpha, \quad \hat{a}_y = \frac{a_y}{|\mathbf{a}|} = \cos\beta, \quad \hat{a}_z = \frac{a_z}{|\mathbf{a}|} = \cos\gamma \cdot$$

This means that $\hat{a}_x$, $\hat{a}_y$, and $\hat{a}_z$ are also the direction cosines of **a**.

Unit Vectors

The following notation convention is useful here: If **p** is any vector, then $\hat{\mathbf{p}}$ denotes its unit vector. The unit vector $\hat{\mathbf{p}}$, in the direction of **p**, is a vector having a magnitude equal to 1; thus, $|\hat{\mathbf{p}}| = 1$. It is given by

$$\hat{\mathbf{p}} = \frac{\mathbf{p}}{|\mathbf{p}|}. \tag{3.15}$$

It follows that

$$\hat{p}_x = \frac{p_x}{|\mathbf{p}|}, \quad \hat{p}_y = \frac{p_y}{|\mathbf{p}|}, \quad \hat{p}_z = \frac{p_z}{|\mathbf{p}|}, \tag{3.16}$$

where $\hat{p}_x$ is the x component of $\hat{\mathbf{p}}$ and

$$|\hat{\mathbf{p}}| = \sqrt{\hat{p}_x^2 + \hat{p}_y^2 + \hat{p}_z^2} = 1, \tag{3.17}$$

or

$$\hat{p}_x^2 + \hat{p}_y^2 + \hat{p}_z^2 = 1. \tag{3.18}$$

Using Equation 3.18 and substituting appropriately produces

$$\hat{p}_x = \cos\alpha, \tag{3.19}$$

$$\hat{p}_y = \cos\beta, \tag{3.20}$$

$$\hat{p}_z = \cos\gamma, \tag{3.21}$$

demonstrating that the components of a unit vector are also its direction cosines.

If $\mathbf{a} = (6\mathbf{i} + 2\mathbf{j} - 5\mathbf{k})$ and $\mathbf{b} = 2\mathbf{a}$, compare their unit vectors $\hat{\mathbf{a}}$ and $\hat{\mathbf{b}}$. We must find that they are equal: $\hat{\mathbf{a}} = \hat{\mathbf{b}}$, because the vectors **a** and **b** differ only by a scale factor. Although the magnitude of **b** is twice that of **a**, they are in the same direction. Therefore, their unit vectors are equal.

The special vectors **i**, **j**, and **k** are unit vectors, where the vector **i** lies along the x axis, **j** along the y axis, and **k** along the z axis, so that **i** = (1,0,0), **j** = (0,1,0), and **k** = (0,0,1), and in matrix format **i** = [1 0 0], **j** = [0 1 0], and **k** = [0 0 1].

Here is how to show that the magnitude of a unit vector is equal to 1: If

$$\hat{\mathbf{a}} = \mathbf{a}/|\mathbf{a}|, \tag{3.22}$$

then

$$|\hat{\mathbf{a}}| = \sqrt{\left(\frac{a_x}{|\mathbf{a}|}\right)^2 + \left(\frac{a_y}{|\mathbf{a}|}\right)^2 + \left(\frac{a_z}{|\mathbf{a}|}\right)^2}, \tag{3.23}$$

or

$$\begin{aligned} |\hat{\mathbf{a}}| &= \sqrt{\frac{a_x^2}{|\mathbf{a}|^2} + \frac{a_y^2}{|\mathbf{a}|^2} + \frac{a_z^2}{|\mathbf{a}|^2}}, \\ &= \frac{\sqrt{a_x^2 + a_y^2 + a_z^2}}{|\mathbf{a}|}, \\ &= 1. \end{aligned}$$

3.4 Addition and Subtraction

Given two vectors $\mathbf{p} = (p_x, p_y, p_z)$ and $\mathbf{q} = (q_x, q_y, q_z)$, to add them we simply add their corresponding components:

$$\mathbf{p} + \mathbf{q} = [(p_x + q_x), (p_y + q_y), (p_z + q_z)], \tag{3.24}$$

or

$$\mathbf{p} + \mathbf{q} = (p_x + q_x)\mathbf{i} + (p_y + q_y)\mathbf{j} + (p_z + q_z)\mathbf{k}. \tag{3.25}$$

The difference of two vectors is

$$\mathbf{p} - \mathbf{q} = [(p_x - q_x), (p_y - q_y), (p_z - q_z)]. \tag{3.26}$$

Order is not important. If

$$\mathbf{d} = \mathbf{a} + \mathbf{b} + \mathbf{c}, \tag{3.27}$$

then it is also true that

$$\mathbf{d} = \mathbf{a} + \mathbf{c} + \mathbf{b}. \tag{3.28}$$

This means that vector addition is commutative.

What is important is that the vectors to be added must represent the same phenomenon. We can use vectors to represent distance, force, velocity, acceleration, and so forth. Adding a distance vector to a force vector is incorrect. Both vectors must be distance, or both must be force . . . you get the picture: apples and apples, oranges and oranges, not apples and oranges.

Here is an example: Given the two-dimensional vectors $\mathbf{p} = (5, 3)$ and $\mathbf{q} = (-2, 5)$ (Figure 3.7), their sum is

$$\begin{aligned} \mathbf{p}+\mathbf{q} &= [(5-2), (3-5)] \\ &= (3, -2), \end{aligned} \tag{3.29}$$

and we see that the parallelogram law of addition is satisfied.

Here is another example: If $\mathbf{a} = (2, 3, 5)$ and $\mathbf{b} = (6, -1, 3)$, then

$$\mathbf{a} + \mathbf{b} = (8, 2, 8), \tag{3.30}$$

$$\mathbf{a} - \mathbf{b} = (-4, 4, 2), \tag{3.31}$$

$$2\mathbf{a} + 3\mathbf{b} = (22, 3, 19). \tag{3.32}$$

Remember that a unit vector's magnitude is equal to 1, independent of its direction. For the unit vector in the direction of **a**, we write $\hat{\mathbf{a}}$, where $\hat{\mathbf{a}} = \mathbf{a}/|\mathbf{a}|$. The components of $\hat{\mathbf{a}}$ are $a_x/|\mathbf{a}|$, $a_y/|\mathbf{a}|$, and $a_z/|\mathbf{a}|$.

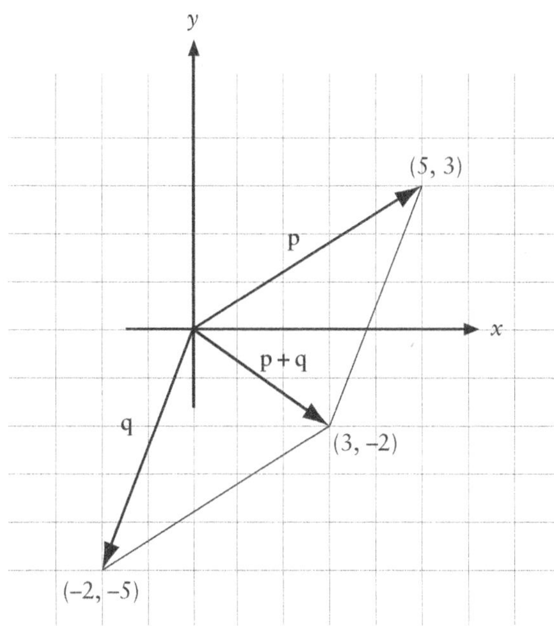

Figure 3.7 Adding two vectors in the plane

3.5 Products of Vectors

This section introduces four different ways to multiply two or three vectors (): the scalar product, the vector product, the scalar triple product, and the vector triple product. Each product yields useful, but quite different, geometric information.

Scalar Product

The scalar product (aka dot product or inner product) of two vectors **p** and **q** is the sum of the products of their corresponding components:

$$\mathbf{p} \bullet \mathbf{q} = p_x q_x + p_y q_y + p_z q_z, \quad (3.33)$$

which turns out to be a scalar. Another way to obtain this value is

$$\mathbf{p} \bullet \mathbf{q} = |\mathbf{p}||\mathbf{q}| \cos\theta, \quad (3.34)$$

so that

$$\theta = \cos^{-1} \frac{\mathbf{p} \bullet \mathbf{q}}{|\mathbf{p}||\mathbf{q}|}, \quad (3.35)$$

where θ is the angle between the two vectors.

Let's practice. Given **a** = [4 −1], **b** = [2 8], **c** = [−4 1], and **d** = [3 2], verify the angles between the following pairs of vectors:

a. **a** and **b** 90°
b. **b** and **c** 180°
c. **c** and **b** 90°
d. **a** and **d** 47.73°
e. **c** and **d** 132.27°

Using the law of cosines, we can demonstrate that the angle θ between two vectors **a** and **b** satisfies the equation

$$\mathbf{a} \bullet \mathbf{b} = |\mathbf{a}||\mathbf{b}| \cos\theta.$$

Solving this for θ yields

$$\theta = \cos^{-1}\left(\frac{\mathbf{a} \bullet \mathbf{b}}{|\mathbf{a}||\mathbf{b}|}\right).$$

This means that if **a** • **b** = 0, then **a** and **b** are perpendicular. If $\theta = 0$, then they are parallel.

This property of the scalar product is immediately useful, because if **p** • **q** = 0, then the two vectors are mutually perpendicular because cos 90° = 0. This also means that the following products are true for the basis vectors **i**, **j**, and **k**:

$$\begin{aligned} \mathbf{i} \bullet \mathbf{i} &= \mathbf{j} \bullet \mathbf{j} = \mathbf{k} \bullet \mathbf{k} = 1, \\ \mathbf{i} \bullet \mathbf{j} &= \mathbf{j} \bullet \mathbf{k} = \mathbf{k} \bullet \mathbf{i} = 0, \\ \mathbf{i} \bullet \mathbf{k} &= \mathbf{j} \bullet \mathbf{i} = \mathbf{k} \bullet \mathbf{j} = 0. \end{aligned} \quad (3.36)$$

Finally, we see that if **p** = **q**, then

$$\mathbf{p} \bullet \mathbf{p} = |\mathbf{p}|^2. \quad (3.37)$$

And last, here is a word about the inner product, which we may think of as an enhanced scalar or dot product. To think of it this way, we get a little outside of this text's plan and mention a more specialized connection to linear vector spaces. The inner product of two vectors is a defining feature of a class of linear vector spaces. It is independent of the coordinate system in which it is expressed, and that includes our familiar Cartesian system in Euclidean space. The inner product operation extracts the length or magnitude of a vector as well as the angle between any two vectors. These are important geometric properties that we obtain from the inner product.

The inner product written in tensor notation is

$$\mathbf{p} \bullet \mathbf{q} = p^j q^k \delta_{jk}, \tag{3.38}$$

where δ_{jk} is the Kronecker delta, and

$$\delta_{jk} = \begin{cases} 1 & \text{if } j = k \\ 0 & \text{if } j \neq k \end{cases}. \tag{3.39}$$

Vector Product

The vector product of two vectors **p** and **q** is another vector, **r**, expressed as

$$\mathbf{p} \times \mathbf{q} = \mathbf{r}, \tag{3.40}$$

which we compute as

$$\mathbf{r} = (p_y q_z - p_z q_y)\mathbf{i} + (p_z q_x - p_x q_z)\mathbf{j} + (p_x q_y - p_y q_x)\mathbf{k}. \tag{3.41}$$

Here is a more explicit statement of the components of **r**:

$$\begin{aligned} r_x &= (p_y q_z - p_z q_y), \\ r_y &= (p_z q_x - p_x q_z), \\ r_z &= (p_x q_y - p_y q_x). \end{aligned} \tag{3.42}$$

The resulting vector **r** has a very important geometric property. It is perpendicular (normal) to both **p** and **q**, and therefore it is perpendicular to the plane defined by them. Among other things, this allows the computation of surface normals, required in many geometric and 3D modeling situations.

The right-hand rule tells us the direction of **r**. Again, think of rotating **p** into **q**, curling the fingers of your right hand in this angular direction. Then the extended thumb of your right hand will point in the direction of **r**.

The expansion of the following determinant also produces the vector product (more about determinants in Chapter 5):

$$\mathbf{p} \times \mathbf{q} = \begin{vmatrix} \mathbf{i} & \mathbf{j} & \mathbf{k} \\ p_x & p_y & p_z \\ q_x & q_y & q_z \end{vmatrix}. \tag{3.43}$$

The vector product of a vector with itself produces the null vector. Using Equation 3.41 we demonstrate this:

$$\begin{aligned}\mathbf{p}\times\mathbf{p} &= (p_y p_z - p_z p_y)\mathbf{i} + (p_z p_x - p_x p_z)\mathbf{j} + (p_x p_y - p_y p_x)\mathbf{k} \\ &= (0, 0, 0)\,.\end{aligned} \tag{3.44}$$

Given $\mathbf{a} = [\,1 \quad 0 \quad -2\,]$, $\mathbf{b} = [\,3 \quad 1 \quad 4\,]$, and $\mathbf{c} = [\,-1 \quad 6 \quad 2\,]$, verify the following vector products:

a. $\mathbf{a} \times \mathbf{a} = 0$
b. $\mathbf{a} \times \mathbf{b} = [\,2 \quad -10 \quad -2\,]$
c. $\mathbf{b} \times \mathbf{a} = [\,-2 \quad 10 \quad 2\,]$
d. $\mathbf{b} \times \mathbf{c} = [\,-22 \quad -10 \quad 19\,]$
e. $\mathbf{c} \times \mathbf{a} = [\,-12 \quad 0 \quad -6\,]$

Now let's look at some geometry revealed by the vector product operation: If $\mathbf{p} \times \mathbf{q} = 0$, then $\mathbf{p}$ and $\mathbf{q}$ are parallel. Furthermore, the angle θ between $\mathbf{p}$ and $\mathbf{q}$ is a by-product of the vector product:

$$|\,\mathbf{p}\times\mathbf{q}\,| = |\,\mathbf{p}\,||\,\mathbf{q}\,|\sin\theta, \tag{3.45}$$

so that

$$\theta = \sin^{-1}\frac{|\,\mathbf{p}\times\mathbf{q}\,|}{|\,\mathbf{p}\,||\,\mathbf{q}\,|}\,. \tag{3.46}$$

We can use this relationship to prove that the area of triangle OAB is given by $|\mathbf{a}\times\mathbf{b}|/2$, where $\mathbf{a}$ is directed line segment OA and $\mathbf{b}$ is OB (Figure 3.8). In the figure we see that the area of the triangle is

$$\text{Area } \Delta OAB = (1/2)|\mathbf{a}||\mathbf{b}|\sin\theta\,. \tag{3.47}$$

Using Equation 3.45, we know that

$$|\mathbf{a}\times\mathbf{b}| = |\mathbf{a}||\mathbf{b}|\sin\theta\,. \tag{3.48}$$

The obvious substitution yields

$$\text{Area } \Delta OAB = \frac{1}{2}|\mathbf{a}\times\mathbf{b}|\,. \tag{3.49}$$

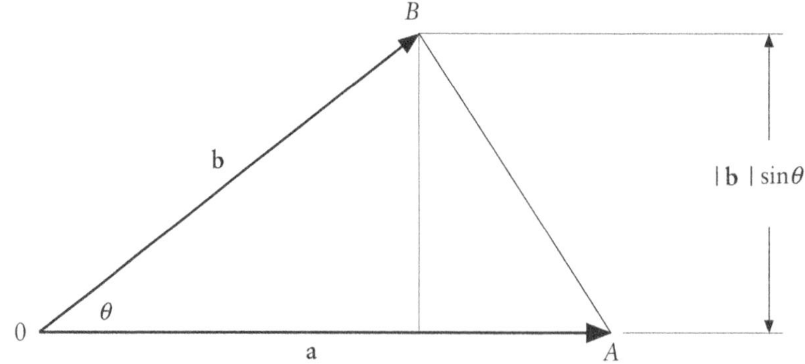

Figure 3.8 Vector product and the area of a triangle

Scalar Triple Product

The scalar triple product **p • q × r** implies **p • (q × r)** and not **(p • q) × r.** (Why?) An easy way to represent how to compute this product is to use the following determinant:

$$\mathbf{p} \bullet \mathbf{q} \times \mathbf{r} = \begin{vmatrix} p_x & q_x & r_x \\ p_y & q_y & r_y \\ p_z & q_z & r_z \end{vmatrix}, \tag{3.50}$$

and it is easy to show that

$$\mathbf{p} \bullet \mathbf{q} \times \mathbf{r} = \mathbf{p} \times \mathbf{q} \bullet \mathbf{r}. \tag{3.51}$$

For those of you who might be interested, a tensor form is

$$\mathbf{p} \bullet \mathbf{q} \times \mathbf{r} = \xi_{ijk} p^i q^j r^k, \tag{3.52}$$

where ξ_{ijk} is the Levi-Cevita symbol, and for the above equation

$$\xi_{ijk} = \begin{Bmatrix} 1 \\ -1 \\ 0 \end{Bmatrix} \text{ if } \mathbf{i}, \mathbf{j}, \mathbf{k} \text{ is an } \begin{Bmatrix} \text{even} \\ \text{odd} \\ \text{other} \end{Bmatrix} \text{ permutation of } 1, 2, 3. \tag{3.53}$$

And what is this about odd and even permutations of the three indices? First, consider combinations (not strictly a permutation) that fall in the "other" category. Examples include 1, 2, 2 and 3, 1, 3. Neither of these two is a permutation of 1, 2, 3. (Notice that we are working with the values the indices can take.) Here is an odd permutation: 1, 3, 2, produced by one (an odd number) exchange . . . 2 and 3 exchanged positions to create a permutation. Here is an even permutation: 2, 3, 1, produced by two exchanges . . . 2 and 3 followed by 1 and 2.

Vector Triple Product

The vector triple product is

$$\begin{aligned} \mathbf{p} \times (\mathbf{q} \times \mathbf{r}) &= (\mathbf{p} \bullet \mathbf{r})\mathbf{q} - (\mathbf{p} \bullet \mathbf{q})\mathbf{r} \\ &= \mathbf{s}, \end{aligned} \tag{3.54}$$

where the result of **p × (q × r)** lies in the plane of **q** and **r.** Try to go through the proof of this.

Here is the tensor form, using the Levi-Cevita form (we won't go into the index protocol, which is much different from that of the scalar triple product):

$$s_i = \xi_{ijk} \xi_{klm} p^j q^l r^m, \tag{3.55}$$

where s_i is the i th component of s.

3.6 Functions and Derivatives

Geometric and 3D modeling depend on the mathematical tools of vector calculus. Some of these are indispensable; others are more useful in special situations. In particular, vector functions and their derivatives and partial derivatives are necessary for even elementary geometric modeling: One-parameter functions define curves, and two-parameter functions define surfaces.

To define a curve in ordinary three-dimensional space, use the vector-valued function $\mathbf{p}(u)$ to define points on it, where u is the independent parametric variable. (Note that u is a scalar variable.) Here are three ways to express this:

$$\mathbf{p}(u) = [\ f_x(u)\mathbf{i} \quad f_y(u)\mathbf{j} \quad f_z(u)\mathbf{k}\ ], \tag{3.56}$$

$$\mathbf{p}(u) = [\ x(u) \quad y(u) \quad z(u)\ ], \tag{3.57}$$

$$\mathbf{p}(u) = f_x(u)\mathbf{i} + f_y(u)\mathbf{j} + f_z(u)\mathbf{k}. \tag{3.58}$$

To define points on a surface, we use vector-valued functions of two independent parametric variables: u and w. Here, again, are three ways to express this:

$$\mathbf{p}(u,w) = [\ f_x(u,w)\mathbf{i} \quad f_y(u,w)\mathbf{j} \quad f_z(u,w)\mathbf{k}\ ], \tag{3.59}$$

$$\mathbf{p}(u,w) = [\ x(u,w) \quad y(u,w) \quad z(u,w)\ ], \tag{3.60}$$

$$\mathbf{p}(u,w) = f_x(u,w)\mathbf{i} + f_y(u,w)\mathbf{j} + f_z(u,w)\mathbf{k}. \tag{3.61}$$

The first derivative of the vector-valued function defining curves is the sum of the derivatives of each of the vector component functions; thus,

$$\frac{d\mathbf{p}(u)}{du} = \frac{df_x(u)}{du}\mathbf{i} + \frac{df_y(u)}{du}\mathbf{j} + \frac{df_z(u)}{du}\mathbf{k}. \tag{3.62}$$

This is the parametric derivative, and as we will see later, it defines the tangent vector at a point $\mathbf{p}(u)$ on the curve.

The relationship between the parametric derivatives and the ordinary derivatives of Cartesian space is

$$\frac{dy}{dx} = \frac{dy/du}{dx/du}, \tag{3.63}$$

and similarly for dy/dz and dz/dx.

We must use partial derivatives when analyzing the vector-valued functions for surfaces. Thus,

$$\frac{\partial \mathbf{p}(u,w)}{\partial u} = \frac{\partial f_x(u,w)}{\partial u}\mathbf{i} + \frac{\partial f_y(u,w)}{\partial u}\mathbf{j} + \frac{\partial f_z(u,w)}{\partial u}\mathbf{k}. \tag{3.64}$$

Now on to a look at the tangent vector to a curve at a point $\mathbf{p}(u_i) = \mathbf{p}_i$ on it. We express this as follows: $\frac{d\mathbf{p}(u)}{du}$ evaluated at $u = u_i$. Another way to indicate this is $\left.\frac{d\mathbf{p}(u)}{du}\right|_{u=u_i}$.

Here is an alternative notation. Its value lies in its simplicity and conciseness. We will denote $\frac{d\mathbf{p}(u)}{du}$ as $\mathbf{p}^u$ and $\left.\frac{d\mathbf{p}(u)}{du}\right|_{u=u_i}$ as $\mathbf{p}_i^u$. Note that denoting $\frac{d\mathbf{p}(u)}{du}$ as $\mathbf{p}^u$ is NOT STANDARD NOTATION, but you will discover that this is a useful time and space saver. Be aware that the superscript u, as on $\mathbf{p}_i^u$, is not an exponent. It instead indicates differentiation with respect to the independent parametric variable u. Notation intent should be clear from the context.

For the second derivative of the function $\mathbf{p}(u)$, we will denote $\frac{d^2\mathbf{p}(u)}{du^2}$ as $\mathbf{p}^{uu}$ and $\left.\frac{d^2\mathbf{p}(u)}{du^2}\right|_{u=u_i}$ as $\mathbf{p}_i^{uu}$. Again, this is NOT STANDARD NOTATION!

Expressions in first- and higher-order derivatives allow us to find geometric properties of curves and surfaces, such as curvature and normals. Later we will see how tangent vectors affect the shape of curves and surfaces.

3.7 Linear Vector Spaces

If we treat vectors as if they originated at a common point, then we work in a linear vector space. Not only do all vectors have a common origin, but also any vector combines with any other vector according to the parallelogram law of addition. Vectors must be subjected to the following two operations to qualify as members of a linear vector space:

1. We must be able to add two vectors to obtain a third vector, identified as their sum: $\mathbf{a} + \mathbf{b} = \mathbf{c}$.
2. We must be able to multiply a vector $\mathbf{a}$ by a scalar k and obtain another vector $k\mathbf{a}$ as the product.

The set of all vectors is closed with respect to these two operations, which means that both the sum of two vectors and the product of a vector and a scalar are themselves vectors. These two operations have the following properties, some of which we have seen before:

- Commutativity: $\mathbf{a} + \mathbf{b} = \mathbf{b} + \mathbf{a}$
- Associativity: $(\mathbf{a} + \mathbf{b}) + \mathbf{c} = \mathbf{a} + (\mathbf{b} + \mathbf{c})$
- Identity element: $\mathbf{a} + 0 = \mathbf{a}$
- Inverse: $\mathbf{a} - \mathbf{a} = 0$

- Identity under scalar multiplication: $k\mathbf{a} = \mathbf{a}$, when $k = 1$
- $c(d\mathbf{a}) = (cd)\mathbf{a}$
- $(c+d)\mathbf{a} = c\mathbf{a} + d\mathbf{a}$
- $k(\mathbf{a}+\mathbf{b}) = k\mathbf{a} + k\mathbf{b}$

A set of vectors that can be subjected to the two operations with these eight properties forms a linear vector space. The other vector operations we have discussed, such as the scalar and vector products, are not pertinent to this definition of a linear vector space. This brings us to spaces of n dimensions. It is easy to show that the set of all vectors of the form $\mathbf{r} = [\ r_1 \quad r_2 \quad \ldots \quad r_n\ ]$, where $r_1, r_2, \ldots, r_n$ are real numbers, constitutes a linear vector in n-dimensional space.

We can form linear combinations of vectors and demonstrate that the set of all linear combinations of a given set forms a vector space. For example, if we let $\mathbf{x}_1, \mathbf{x}_2, \ldots, \mathbf{x}_n$ be any n vectors, then $a_1\mathbf{x}_1 + a_2\mathbf{x}_2 + \ldots + a_n\mathbf{x}_n$ (where $a_1, a_2, \ldots, a_n$ are scalars) is a linear combination of the vectors $\mathbf{x}_1, \mathbf{x}_2, \ldots \mathbf{x}_n$. That is not all we can do. If we let

$$\begin{aligned} \mathbf{s} &= a_1\mathbf{x}_1 + a_2\mathbf{x}_2 + \ldots + a_n\mathbf{x}_n, \\ \mathbf{t} &= b_1\mathbf{x}_1 + b_2\mathbf{x}_2 + \ldots + b_n\mathbf{x}_n, \end{aligned} \tag{3.65}$$

so that the vectors $\mathbf{s}$ and $\mathbf{t}$ are linear combinations of the vectors $\mathbf{x}_1, \mathbf{x}_2, \ldots, \mathbf{x}_n$, then

$$\mathbf{s} + \mathbf{t} = (a_1 + b_1)\mathbf{x}_1 + (a_2 + b_2)\mathbf{x}_2 + \ldots + (a_n + b_n)\mathbf{x}_n. \tag{3.66}$$

We also have

$$k\mathbf{s} = (ka_1)\mathbf{x}_1 + (ka_2)\mathbf{x}_2 + \ldots + (ka_n)\mathbf{x}_n, \tag{3.67}$$

which is also a linear combination of $\mathbf{x}_1, \mathbf{x}_2, \ldots, \mathbf{x}_n$. Mathematicians point out that the space of all linear combinations of a given set of vectors is the space generated by that set.

Given a single vector $\mathbf{x}$, the space generated by all scalar multiples of $\mathbf{x}$ is a straight line collinear with $\mathbf{x}$. Given two vectors $\mathbf{s}$ and $\mathbf{t}$, where $\mathbf{t}$ is not a scalar multiple of $\mathbf{s}$, then the space generated by all their linear combinations is the plane containing $\mathbf{s}$ and $\mathbf{t}$. For example, if we let $\mathbf{r} = a\mathbf{s} + b\mathbf{t}$, then from the parallelogram law of addition, we know that vectors $\mathbf{r}$, $\mathbf{s}$, and $\mathbf{t}$ are coplanar. Of course, we could continue this process, generating spaces of three and more dimensions simply by increasing the number of vectors in the generating set. To do this, we must impose certain conditions, as in the previous example where we did not allow $\mathbf{s}$ and $\mathbf{t}$ to be scalar multiples of each other. This leads us to the concepts of linear independence and dependence.

Two vectors that point in different directions are linearly independent. This is the simplest and most graphical or visual way to define the concept of their linear independence. Two vectors are linearly dependent if they differ only by a scalar value. For example, if $\mathbf{a} = k\mathbf{b}$, then $\mathbf{a}$ and $\mathbf{b}$ are linearly dependent. However, there is a more computationally useful definition. Vectors $\mathbf{x}_1, \mathbf{x}_2, \ldots, \mathbf{x}_n$ are linearly dependent if and only if there are real numbers $a_1, a_2, \ldots, a_n$ not all equal to zero, such that

$$a_1\mathbf{x}_1 + a_2\mathbf{x}_2 + \ldots + a_n\mathbf{x}_n = 0. \tag{3.68}$$

If this equation is true only if $a_1, a_2, \ldots, a_n$ are all zero, then $\mathbf{x}_1, \mathbf{x}_1, \ldots, \mathbf{x}_n$ are *linearly independent.*

If $\mathbf{x}_1, \mathbf{x}_1, \ldots, \mathbf{x}_n$ are linearly dependent, then it is possible to express any one of them as a linear combination of the others. On the other hand, if one of the vectors $\mathbf{x}_1, \mathbf{x}_2, \ldots, \mathbf{x}_n$ is a linear combination of the others, then the vectors are linearly dependent. Another way of saying this is that vectors $\mathbf{x}_1, \mathbf{x}_2, \ldots, \mathbf{x}_n$ are linearly dependent if and only if one of them belongs to the space generated by the remaining $n - 1$ vectors. The dimension of a linear space is equal to the maximum number of linearly independent vectors that it can contain. This fact underlies the study of basis vectors, introduced in Chapter 6.

Any two vectors are dependent if and only if they are parallel (or collinear); three vectors are dependent if and only if they are coplanar; four vectors are dependent in a space of three dimensions; n vectors are dependent in a space of $n - 1$ dimensions. Finally, in a space of three dimensions, a set of three vectors $\mathbf{r}$, $\mathbf{s}$, and $\mathbf{t}$ is linearly dependent if and only if the determinant

$$\begin{vmatrix} r_x & r_y & r_z \\ s_x & s_y & s_z \\ t_x & t_y & t_z \end{vmatrix} = 0. \tag{3.69}$$

If the vectors $\mathbf{x}_1, \mathbf{x}_1, \ldots, \mathbf{x}_n$ are linearly independent, it is impossible to represent any one of them as a linear combination of the other $n - 1$ vectors.

3.8 Vector Equations

. . . Or how to solve math problems using vectors. For example, given that the vector equation $\mathbf{a} + u\mathbf{b} + w\mathbf{c} = \mathbf{d} + t\mathbf{e}$ represents a system of three linear equations in three unknowns, solve for u, w, and v. In component form these equations are

$$\begin{aligned} a_x + ub_x + wc_x &= d_x + te_x\,, \\ a_y + ub_y + wc_y &= d_y + te_y\,, \\ a_z + ub_z + wc_z &= d_z + te_z\,. \end{aligned} \tag{3.70}$$

First, to isolate t, apply $(\mathbf{b} \times \mathbf{c})$ as follows:

$$(\mathbf{b} \times \mathbf{c}) \bullet (\mathbf{a} + u\mathbf{b} + w\mathbf{c}) = (\mathbf{b} \times \mathbf{c}) \bullet (\mathbf{d} + t\mathbf{e}). \tag{3.71}$$

Because $(\mathbf{b} \times \mathbf{c})$ is perpendicular to both $\mathbf{b}$ and $\mathbf{c}$, then

$$(\mathbf{b}\times\mathbf{c})\bullet\mathbf{a}=(\mathbf{b}\times\mathbf{c})\bullet\mathbf{d}+(t\mathbf{b}\times\mathbf{c})\bullet\mathbf{e}. \tag{3.72}$$

Solving this equation for t produces

$$t=\frac{(\mathbf{b}\times\mathbf{c})\bullet\mathbf{a}-(\mathbf{b}\times\mathbf{c})\bullet\mathbf{d}}{(\mathbf{b}\times\mathbf{c})\bullet\mathbf{e}}. \tag{3.73}$$

Continuing this method yields similar expressions for u and w:

$$\begin{aligned} u&=\frac{(\mathbf{c}\times\mathbf{e})\bullet\mathbf{d}-(\mathbf{c}\times\mathbf{e})\bullet\mathbf{a}}{(\mathbf{c}\times\mathbf{e})\bullet\mathbf{b}},\\ w&=\frac{(\mathbf{b}\times\mathbf{e})\bullet\mathbf{d}-(\mathbf{b}\times\mathbf{e})\bullet\mathbf{a}}{(\mathbf{b}\times\mathbf{e})\bullet\mathbf{c}}. \end{aligned} \tag{3.74}$$

3.9 Quaternions

Now may be the time to introduce quaternions. Quaternions are an extension of complex numbers and vectors, having applications in both theoretical and applied mathematics, physics, and engineering. They are a particularly good tool to use to compute the product of sequential rotations about arbitrary axes and to extract the equivalent rotation angle and axis, including calculations of rotations for 3D animation. Let's start with a review of complex numbers.

A complex number has two parts, a real part and an imaginary part. Here is an example:

$$c = a + bi, \tag{3.75}$$

where $i^2 = -1$, a is the real part, and b is the imaginary part, albeit both a and b are real numbers. The conjugate of c is c^*, where

$$c^* = a - bi. \tag{3.76}$$

A geometric interpretation of complex numbers reveals itself as points on the complex plane, a construction in a two-dimensional coordinate system with the real numbers, a, measured on the horizontal axis, and the imaginary numbers, b, measured on the vertical axis (Figure 3.9).

Here are some examples of operations on complex numbers. Try to confirm the results.

a. $(3+2i)+(1-4i)=(4-2i)$
b. $(7-i)+(1+3i)=(8+2i)$
c. $4i+(-3+2i)=(-3+6i)$
d. $(6-5i)\times i=(5+6i)$
e. $(a+bi)\times(a-bi)=a^2+b^2$

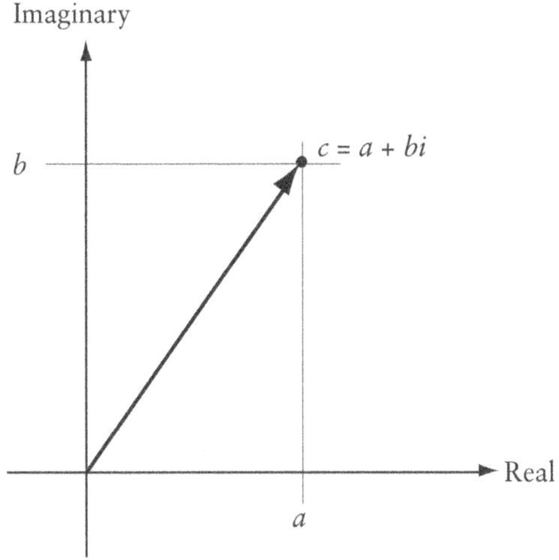

Figure 3.9 The complex plane

And here are some rules to keep in mind:

1. Two complex numbers are equal if and only if their corresponding real parts are equal and their imaginary parts are equal.
2. Given two complex numbers c_1 and c_2, where

$$c_1 = a_1 + b_1 i \tag{3.77}$$

and

$$c_2 = a_2 + b_2 i, \tag{3.78}$$

their sum is

$$c_1 + c_2 = a_1 + a_2 + (b_1 + b_2)i. \tag{3.79}$$

3. The product of two complex numbers c_1 and c_2 is

$$\begin{aligned} c_1 c_2 &= (a_1 + b_1 i)(a_2 + b_2 i) \\ &= a_1 a_2 + a_1 b_2 i + a_2 b_1 i + b_1 b_2 i^2, \end{aligned} \tag{3.80}$$

or

$$c_1 c_2 = (a_1 a_2 - b_1 b_2) + (a_1 b_2 + a_2 b_1)i. \tag{3.81}$$

It is easy to show that the multiplication of two complex numbers is commutative, meaning

$$c_1 c_2 = c_2 c_1. \tag{3.82}$$

Quaternion math expands on that of complex numbers, defining a new kind of number or mathematical object, Q, with special operating rules and interpretations. Where complex numbers have two parts, quaternions have four parts. This is how we define and represent a quaternion:

$$Q = a + bi + cj + dk, \tag{3.83}$$

where a, b, c, and d are real numbers and i, j, and k are quaternion units, with the following properties:

$$i = j = k = \sqrt{-1}, \tag{3.84}$$

and

$$i^2 = j^2 = k^2 = ijk = -1, \tag{3.85}$$

and also

$$\begin{matrix} ij = k & ji = -k \\ jk = i & kj = -i \\ ki = j & ik = -j \end{matrix} . \tag{3.86}$$

The terms i, j, and k are similar to the unit basis vectors **i**, **j**, and **k** of an orthogonal Cartesian coordinate system, with the additional properties stated by Equations 3.85 and 3.86. Compare the quaternion unit properties with those of the unit basis vectors, where

$$\begin{matrix} \mathbf{i} \bullet \mathbf{i} = \mathbf{j} \bullet \mathbf{j} = \mathbf{k} \bullet \mathbf{k} = 1 & & \mathbf{i} \times \mathbf{j} = \mathbf{k} \\ \mathbf{i} \bullet \mathbf{j} = \mathbf{j} \bullet \mathbf{k} = \mathbf{k} \bullet \mathbf{i} = 0 & \text{and} & \mathbf{j} \times \mathbf{k} = \mathbf{i} \\ \mathbf{i} \bullet \mathbf{k} = \mathbf{j} \bullet \mathbf{i} = \mathbf{k} \bullet \mathbf{j} = 0 & & \mathbf{k} \times \mathbf{i} = \mathbf{j}. \end{matrix} \tag{3.87}$$

And, of course, unit basis vectors are not complex numbers.

This text uses regular lowercase italic for the quaternion units to make a notational distinction from the unit basis vectors. Mathematicians treat a quaternion as equivalent to a vector if the first term of the quaternion equals zero. So, again, given the quaternion

$$Q = a + bi + cj + dk, \tag{3.88}$$

if $a = 0$, then we associate Q with a three-dimensional vector **q**, where i, j, and k are now interpreted as unit basis vectors along the coordinate axes, that is, **i**, **j**, and **k**.

The set of quaternions for which b, c, and d are all equal to zero is equivalent to the set of real numbers. The set of quaternions for which only one of b, c, and d is nonzero is equivalent to the set of complex numbers of the form $z = x + iy$.

Thus, the quaternion

$$Q = 0 + xi + yj + zk \tag{3.89}$$

represents the vector

$$\mathbf{q} = x\mathbf{i} + y\mathbf{j} + z\mathbf{k}. \tag{3.90}$$

Furthermore, we see that

$$Q = a + m\mathbf{q}, \tag{3.91}$$

or

$$Q = a + mxi + myj + mzk. \tag{3.92}$$

The conjugate of Q is Q^*, so that

$$Q^* = a - bi - cj - dk. \tag{3.93}$$

Now let's review some quaternion algebra that will lead to their connection with rotations. First, the sum of two quaternions $Q_1 + Q_2$, where $Q_1 = a_1 + \mathbf{r}$ and $Q_2 = a_2 + \mathbf{s}$, is another quaternion such that

$$Q_1 + Q_2 = (a_1 + a_2) + (\mathbf{r} + \mathbf{s}). \tag{3.94}$$

Here is another format for quaternion addition:

$$Q_1 = a_1 + b_1 i + c_1 j + d_1 k, \tag{3.95}$$

$$Q_2 = a_2 + b_2 i + c_2 j + d_2 k, \tag{3.96}$$

and so

$$Q_1 + Q_2 = (a_1 + a_2) + (b_1 + b_2)i + (c_1 + c_2)j + (d_1 + d_2)k. \tag{3.97}$$

The product of a scalar and a quaternion is

$$mQ = ma + mbi + mcj + mdk. \tag{3.98}$$

The product of two quaternions is

$$\begin{aligned} Q_1 Q_2 &= (a_1 + \mathbf{r})(a_2 + \mathbf{s}) \\ &= (a_1 a_2 - \mathbf{r} \bullet \mathbf{s}) + (a_1 \mathbf{s} + a_2 \mathbf{r} + \mathbf{r} \times \mathbf{s}), \end{aligned} \tag{3.99}$$

or in expanded format

$$\begin{aligned} Q_1 Q_2 &= (a_1 + b_1 i + c_1 j + d_1 k)(a_2 + b_2 i + c_2 j + d_2 k) \\ &= a_1 a_2 + a_1 b_2 i + a_1 c_2 j + a_1 d_2 k \\ &\quad + b_1 a_2 i + b_1 b_2 i^2 + b_1 c_2 ij + b_1 d_2 ik \\ &\quad + c_1 a_2 j + c_1 b_2 ij + c_1 c_2 j^2 + c_1 d_2 jk \\ &\quad + d_1 a_2 k + d_1 b_2 ik + d_1 c_2 jk + d_1 d_2 k^2. \end{aligned} \tag{3.100}$$

After rearranging and computing quaternion unit products, we find

$$\begin{aligned} Q_1 Q_2 &= a_1 a_2 - b_1 b_2 - c_1 c_2 - d_1 d_2 \\ &\quad + (a_1 b_2 + b_1 a_2 + c_1 d_2 - d_1 c_2)i \\ &\quad + (a_1 c_2 - b_1 d_2 + c_1 a_2 + d_1 b_2)j \\ &\quad + (a_1 d_2 + b_1 c_2 - c_1 b_2 + d_1 a_2)k. \end{aligned} \tag{3.101}$$

The result in Equation 3.99 consists of a scalar part and a vector part. The presence of the term $\mathbf{r} \times \mathbf{s}$ in the vector part of the quaternion indicates that the vector product is not commutative; that is

$$Q_1Q_2 \neq Q_2Q_1. \tag{3.102}$$

The inner product looks like this:

$$\begin{aligned} Q_1 \bullet Q_2 &= (a_1, \mathbf{r}) \bullet (a_2, \mathbf{s}) \\ &= a_1a_2 + r_1s_1 + r_2s_2 + r_3s_3 \\ &= a_1a_2 + \mathbf{r} \bullet \mathbf{s}. \end{aligned} \tag{3.103}$$

The scalar product of the vector components is

$$Q_1 \bullet Q_2 = b_1b_2 + c_1c_2 + d_1d_2. \tag{3.104}$$

The vector product of the vector components is

$$Q_1 \times Q_2 = (c_1d_2 - d_1c_2)\mathbf{i} + (d_1b_2 - b_1d_2)\mathbf{j} + (b_1c_2 - c_1b_2)\mathbf{k}. \tag{3.105}$$

Finally, the unit length quaternion must satisfy

$$\begin{aligned} Q_1 \bullet Q_2 &= a^2 + \mathbf{r} \bullet \mathbf{r} \\ &= 1. \end{aligned} \tag{3.106}$$

3.10 Geometry Problems

Let's use vectors to solve some elementary geometry properties.

Problem 1: The Law of Sines

From elementary trigonometry, the law of sines states

$$\frac{a}{\sin A} = \frac{b}{\sin B} = \frac{c}{\sin C},$$

where triangle ABC has sides of length $a = |\mathbf{a}|$, $b = |\mathbf{b}|$, and $c = |\mathbf{c}|$ (Figure 3.10). We will use vector products to prove this.

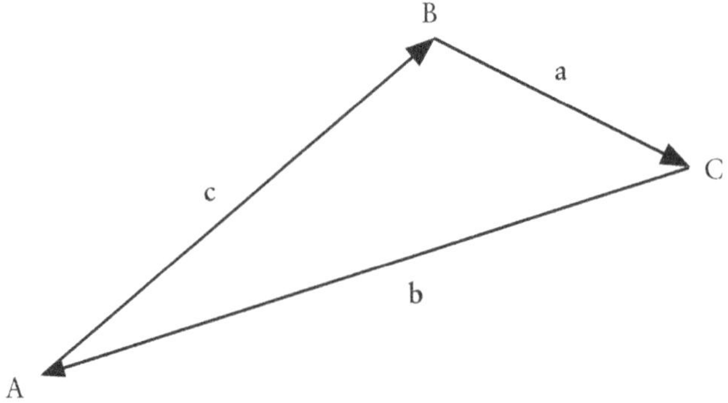

Figure 3.10 Problem 1: The law of sines

Begin by denoting each side of the triangle as a vector. Observe that $\overrightarrow{AB} + \overrightarrow{BC} + \overrightarrow{CA} = 0$. Take the vector products of this expression with each vector:

$$\left(\overrightarrow{AB}+\overrightarrow{BC}+\overrightarrow{CA}\right)\times\overrightarrow{AB}=0$$
$$\left(\overrightarrow{AB}+\overrightarrow{BC}+\overrightarrow{CA}\right)\times\overrightarrow{BC}=0$$
$$\left(\overrightarrow{AB}+\overrightarrow{BC}+\overrightarrow{CA}\right)\times\overrightarrow{AC}=0.$$

This yields

$$\begin{aligned}\overrightarrow{BC}\times\overrightarrow{AB}&=\overrightarrow{AB}\times\overrightarrow{CA}\quad\text{or}\quad ac\sin B=bc\sin A\\ \overrightarrow{AB}\times\overrightarrow{BC}&=\overrightarrow{BC}\times\overrightarrow{CA}\quad\text{or}\quad ac\sin B=ab\sin C\\ \overrightarrow{AB}\times\overrightarrow{CA}&=\overrightarrow{CA}\times\overrightarrow{BC}\quad\text{or}\quad bc\sin A=ab\sin C,\end{aligned}$$

and thus $\frac{a}{\sin A}=\frac{b}{\sin B}=\frac{c}{\sin C}$, Q.E.D.

Problem 2: The Area of a Triangle

Earlier we saw that the area of triangle OAB is given by $|\mathbf{a}\times\mathbf{b}|/2$, where **a** is directed line segment OA and **b** is OB (Figure 3.11).

The area of ΔOAB is $\Delta OAB=\frac{1}{2}|\mathbf{a}||\mathbf{b}|\sin\theta$.

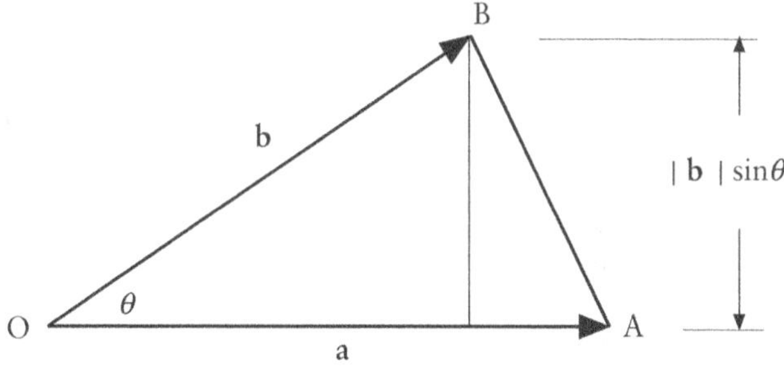

Figure 3.11 Problem 2: The area of a triangle

We use the relationship $|\mathbf{a}\times\mathbf{b}|=|\mathbf{a}||\mathbf{b}|\sin\theta$, perform the obvious substitution, and find that the area is Area $\Delta OAB=\frac{1}{2}|\mathbf{a}\times\mathbf{b}|$.

Show that this works for any orientation in space.

Problem 3: A Property of Triangles

Given a triangle in three-dimensional Cartesian space with sides **a**, **b**, and **a** + **b**, the inequality $|\mathbf{a}+\mathbf{b}|\le|\mathbf{a}|+|\mathbf{b}|$ demonstrates that the sum of the lengths of any two sides of a triangle is greater than the length of the third side.

Here is how we prove this:

$$|\mathbf{a}+\mathbf{b}|^2 = (\mathbf{a}+\mathbf{b})\bullet(\mathbf{a}+\mathbf{b})$$
$$= |\mathbf{a}|^2 + |\mathbf{b}|^2 + 2(\mathbf{a}\bullet\mathbf{b}) \le (|\mathbf{a}|+|\mathbf{b}|)^2$$
$$= |\mathbf{a}|^2 + |\mathbf{b}|^2 + 2(\mathbf{a}\bullet\mathbf{b}) \le |\mathbf{a}|^2 + |\mathbf{b}|^2 + 2|\mathbf{a}||\mathbf{b}|$$

iff $\mathbf{a}\bullet\mathbf{b} \le |\mathbf{a}||\mathbf{b}|$

but $\mathbf{a}\bullet\mathbf{b} \le |\mathbf{a}||\mathbf{b}|\cos\theta$

if $\theta > 0°$, then $\mathbf{a}\bullet\mathbf{b} \le |\mathbf{a}||\mathbf{b}|$

Q.E.D.

3.11 Exercises

3.1. Given the five vectors shown in Figure 3.12, write them in component form. The figure is an orthogonal grid shown at unit intervals.

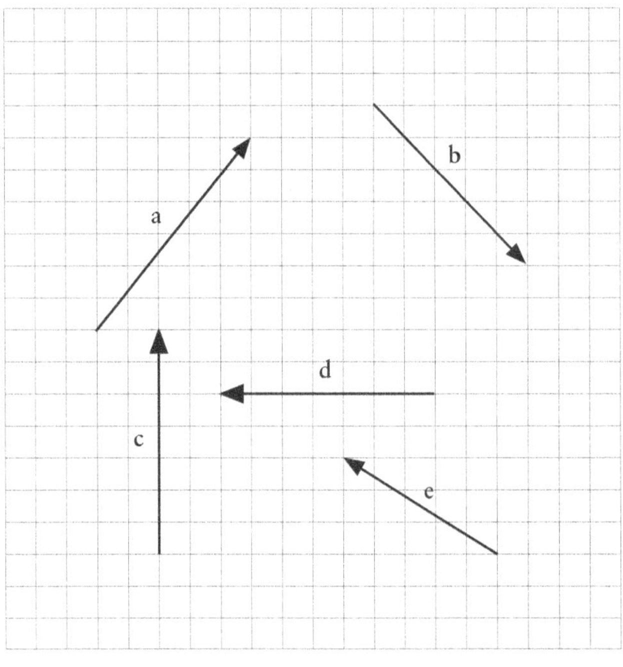

Figure 3.12 Vectors on a grid

3.2. Compute the magnitudes of the vectors given in Exercise 3.1.

3.3. Given that **a**, **b**, and **c** are three-component vectors, express the compact vector equation $\mathbf{a}x + \mathbf{b}y = \mathbf{c}$

a. In expanded vector form.

b. In ordinary algebraic polynomial form.

3.4. Given $\mathbf{a} = [\,-2 \quad 0 \quad 7\,]$ and $\mathbf{b} = [\,4 \quad 1 \quad 3\,]$, compute

a. $\hat{\mathbf{a}}$

b. $\hat{\mathbf{b}}$

c. $\mathbf{c} = \mathbf{a} - 2\mathbf{b}$

d. $\mathbf{c} = 3\mathbf{a}$

e. $\mathbf{c} = \mathbf{a} + \mathbf{b}$

3.5. Given $\mathbf{a} = [\,6 \quad 2 \quad -5\,]$ and $\mathbf{b} = 2\mathbf{a}$, compare the unit vectors $\hat{\mathbf{a}}$ and $\hat{\mathbf{b}}$.

3.6. Given $\mathbf{a} = [\,2 \quad 3 \quad 5\,]$ and $\mathbf{b} = [\,6 \quad -1 \quad 3\,]$, compute

a. $|\mathbf{a}|$

b. $|\mathbf{b}|$

c. $\mathbf{c} = \mathbf{a} + \mathbf{b}$

d. $\mathbf{c} = \mathbf{a} - \mathbf{b}$

e. $\mathbf{c} = 2\mathbf{a} + 3\mathbf{b}$

3.7. Compute the following scalar products:

a. $\mathbf{i} \bullet \mathbf{i}$

b. $\mathbf{i} \bullet \mathbf{j}$

c. $\mathbf{i} \bullet \mathbf{k}$

d. $\mathbf{j} \bullet \mathbf{j}$

e. $\mathbf{j} \bullet \mathbf{k}$

f. $\mathbf{k} \bullet \mathbf{k}$

3.8. Compute the magnitude and direction cosines for each of the following vectors:

a. $\mathbf{a} = [\,3 \quad 4\,]$

b. $\mathbf{b} = [\,0 \quad -2\,]$

c. $\mathbf{c} = [\,-3 \quad -5 \quad 0\,]$

d. $\mathbf{d} = [\,1 \quad 4 \quad -3\,]$

e. $\mathbf{e} = [\,x \quad y \quad z\,]$

3.9. Compute the scalar product of the following pairs of vectors:

a. $\mathbf{a} = [\,0 \quad -2\,]$, $\mathbf{b} = [\,1 \quad 3\,]$

b. $\mathbf{a} = [\,4 \quad -1\,]$, $\mathbf{b} = [\,2 \quad 1\,]$

c. $\mathbf{a} = [\,1 \quad 0\,]$, $\mathbf{b} = [\,0 \quad 4\,]$

d. $\mathbf{a} = [\,3 \quad 0 \quad -2\,]$, $\mathbf{b} = [\,0 \quad -1 \quad -3\,]$
e. $\mathbf{a} = [\,5 \quad 1 \quad 7\,]$, $\mathbf{b} = [\,-2 \quad 4 \quad 1\,]$

3.10. Compute the following vector products:

a. $\mathbf{i} \times \mathbf{i}$
b. $\mathbf{j} \times \mathbf{j}$
c. $\mathbf{k} \times \mathbf{k}$
d. $\mathbf{i} \times \mathbf{j}$
e. $\mathbf{j} \times \mathbf{k}$
f. $\mathbf{k} \times \mathbf{i}$
g. $\mathbf{j} \times \mathbf{i}$
h. $\mathbf{k} \times \mathbf{j}$
i. $\mathbf{i} \times \mathbf{k}$

3.11. Show that $\mathbf{b} \times \mathbf{a} = -(\mathbf{a} \times \mathbf{b})$.
3.12. Compute the angle between the following pairs of vectors:

a. $\mathbf{a} = [\,0 \quad -2\,]$, $\mathbf{b} = [\,1 \quad 3\,]$
b. $\mathbf{a} = [\,4 \quad -1\,]$, $\mathbf{b} = [\,2 \quad 1\,]$
c. $\mathbf{a} = [\,1 \quad 0\,]$, $\mathbf{b} = [\,0 \quad 4\,]$
d. $\mathbf{a} = [\,3 \quad 0 \quad -2\,]$, $\mathbf{b} = [\,0 \quad -1 \quad -3\,]$
e. $\mathbf{a} = [\,5 \quad 1 \quad 7\,]$, $\mathbf{b} = [\,-2 \quad 4 \quad 1\,]$

3.13. Compute the vector product for each of the following pairs of vectors:

a. $\mathbf{a} = [\,3 \quad -1 \quad 2\,]$, $\mathbf{b} = [\,2 \quad 0 \quad 2\,]$
b. $\mathbf{b} = [\,4 \quad 1 \quad -5\,]$, $\mathbf{b} = [\,3 \quad 6 \quad 2\,]$
c. $\mathbf{a} = [\,2 \quad -1 \quad 3\,]$, $\mathbf{b} = [\,-4 \quad 2 \quad -6\,]$
d. $\mathbf{a} = [\,0 \quad 1 \quad 0\,]$, $\mathbf{b} = [\,1 \quad 0 \quad 0\,]$
e. $\mathbf{a} = [\,0 \quad 0 \quad 1\,]$, $\mathbf{b} = [\,1 \quad 0 \quad 0\,]$

3.14. Show that vectors $\mathbf{a} = \begin{bmatrix} -\frac{1}{3} & \frac{2}{3} & \frac{2}{3} \end{bmatrix}$, $\mathbf{b} = \begin{bmatrix} \frac{2}{3} & -\frac{1}{3} & \frac{2}{3} \end{bmatrix}$, and $\mathbf{c} = \begin{bmatrix} -\frac{2}{3} & -\frac{2}{3} & \frac{1}{3} \end{bmatrix}$ are mutually perpendicular.

3.15. Show that the line joining the midpoints of two sides of a triangle is parallel to the third side and has one-half its magnitude.

3.16. Is the set of all vectors lying in the first quadrant of the xy plane a linear space? Why?

3.17. Determine nontrivial linear relations for the following sets of vectors:

a. $\mathbf{p} = [\ 1\ \ 0\ \ -2\ ]$, $\mathbf{q} = [\ 3\ \ -1\ \ 3\ ]$, $\mathbf{r} = [\ 5\ \ -2\ \ 8\ ]$

b. $\mathbf{p} = [\ 2\ \ 0\ \ 1\ ]$, $\mathbf{q} = [\ 0\ \ 5\ \ 1\ ]$, $\mathbf{r} = [\ 6\ \ -5\ \ 4\ ]$

c. $\mathbf{p} = [\ 3\ \ 0\ ]$, $\mathbf{q} = [\ 1\ \ 4\ ]$, $\mathbf{r} = [\ 2\ \ -1\ ]$

3.18. Are the vectors $\mathbf{r} = [\ 1\ \ -1\ \ 0\ ]$, $\mathbf{s} = [\ 0\ \ 2\ \ -1\ ]$, and $\mathbf{t} = [\ 2\ \ 0\ \ -1\ ]$ linearly dependent? Why?

3.19. Take the vector product of two vectors **a** and **b** in two dimensions. The resulting vector has a magnitude $ab\sin\theta$. What about its direction? Where does it point? Remember, this is a two-dimensional problem.

3.20. Repeat Exercise 3.19 for the vector product in four dimensions.

3.21. Show that $(\mathbf{p}-\mathbf{q})\bullet(\mathbf{p}+\mathbf{q}) = |\mathbf{p}|^2 - |\mathbf{q}|^2$ and that $(\mathbf{p}-\mathbf{q})\times(\mathbf{p}+\mathbf{q}) = 2\mathbf{p}\times\mathbf{q}$. Interpret the results with an appropriate sketch.

3.22. Given vectors $\mathbf{a} = 6\mathbf{i} + 10\mathbf{j} + 2\mathbf{k}$ and $\mathbf{b} = \mathbf{i} + 2\mathbf{j} + 6\mathbf{k}$, find $\mathbf{c} = \mathbf{a} + \mathbf{b}$ (**i**, **j**, and **k** are unit vectors in the x, y, and z directions, respectively).

3.23. Given that **p** and **q** are linearly independent vectors in the plane, find the value of k that makes each of the following pairs of vectors collinear:

a. $k\mathbf{p} + 2\mathbf{q}$, $\mathbf{p} - \mathbf{q}$

b. $(k + 1)\mathbf{p} + \mathbf{q}$, $2\mathbf{q}$

c. $k\mathbf{p} + \mathbf{q}$, $\mathbf{p} + k\mathbf{q}$

3.24. Show that the magnitude of a unit vector is equal to 1.

3.25. Find the magnitude and direction numbers for each of the following vectors:

a. $\mathbf{a} = (3, 4)$

b. $\mathbf{b} = (0, -2)$

c. $\mathbf{c} = (-3, -5, 0)$

d. $\mathbf{d} = (1, 4, -3)$

e. $\mathbf{e} = (x, y, z)$

3.26. Find the inner product of the following pairs of vectors:

a. $[0\ \ -2]$, $[1\ \ 3]$
b. $[4\ \ -1]$, $[2\ \ 1]$
c. $[1\ \ 0]$, $[0\ \ 4]$
d. $[3\ \ 0\ \ -2]$, $[0\ \ -1\ \ -3]$
e. $[5\ \ 1\ \ 7]$, $[-2\ \ 4\ \ 1]$

3.27. Find the angle between each pair of vectors given in Exercise 3.26.

3.28. Find the vector product for each of the following pairs of vectors:

a. $[3\ \ -1\ \ 2]$, $[2\ \ 0\ \ 2]$
b. $[4\ \ 1\ \ -5]$, $[3\ \ 6\ \ 2]$
c. $[2\ \ -1\ \ 3]$, $[-4\ \ 2\ \ -6]$
d. $[0\ \ 1\ \ 0]$, $[1\ \ 0\ \ 0]$
e. $[0\ \ 0\ \ 1]$, $[1\ \ 0\ \ 0]$

4 Matrix Basics

A matrix is a way to organize and process certain kinds of mathematical data. The coefficients of a set of linear equations form a matrix. The coordinates of a point suitably arranged also form a matrix. Matrix algebra allows one matrix to operate on another to perform computations more efficiently. Matrices help to solve complex systems of equations, to execute geometric transformations, such as translation, rotation, and scaling, and to represent geometric objects in a computer database. Matrix algebra defines these operations. Vectors and matrices support each other and, to some extent, tensors too. This chapter covers some of the basics.

4.1 Equality and Order

A matrix is a rectangular array of numbers or their algebraic equivalents, arranged in m rows and n columns. We denote a matrix with a boldface uppercase letter, such as **A**, **B**, **C**, **P**, . . ., **T**, and use brackets to enclose the array. For example,

$$\mathbf{A} = \begin{bmatrix} a_{11} & a_{12} & a_{13} \\ a_{21} & a_{22} & a_{23} \\ a_{31} & a_{32} & a_{33} \\ a_{41} & a_{42} & a_{43} \end{bmatrix}. \tag{4.1}$$

The lowercase letters are the elements of the matrix. Each element has a double subscript that gives the position of the element in the array by row and column number. For example, a_{32} is in the third row and second column, and a_{ij} is in row i and column j. A comma is used between subscript indices or numbers only when there are 10 or more rows or columns. The number of rows and columns defines the order of a matrix. The order of the matrix above is 4×3 (pronounced "four by three").

Two matrices are equal if they are the same order and their corresponding elements are equal. No equality occurs between matrices of different orders. Here are three matrices (**A**, **B**, and **C**):

$$\mathbf{A} = \begin{bmatrix} a_{11} & a_{12} \\ a_{21} & a_{22} \\ a_{31} & a_{32} \end{bmatrix},$$
$$\mathbf{B} = \begin{bmatrix} b_{11} & b_{12} \\ b_{21} & b_{22} \\ b_{31} & b_{32} \end{bmatrix}, \tag{4.2}$$
$$\mathbf{C} = \begin{bmatrix} c_{11} & c_{12} & c_{13} \\ c_{21} & c_{22} & c_{23} \\ c_{31} & c_{32} & c_{33} \end{bmatrix}.$$

Matrices **A** and **B** are the same order, so **A** = **B** if and only if $a_{ij} = b_{ij}$ for $i = 1, 2, 3$ and $j = 1, 2$. This means that **A** = **B** if and only if $a_{11} = b_{11}$, $a_{12} = b_{12}$, $a_{21} = b_{21}$, $a_{22} = b_{22}$, $a_{31} = b_{31}$, and $a_{32} = b_{32}$. Obviously, **A** ≠ **C** and **B** ≠ **C** because their elements cannot be equated one to one. An $m \times n$ matrix cannot equal an $n \times m$ matrix if $m \neq n$. Even though they have the same number of elements, they are not the same order.

For any given matrix, if the number of rows equals the number of columns, $m = n$, then it is a square matrix (more on this in the next chapter). For example, **C** is a 3×3 square matrix:

$$\mathbf{C} = \begin{bmatrix} c_{11} & c_{12} & c_{13} \\ c_{21} & c_{22} & c_{23} \\ c_{31} & c_{32} & c_{33} \end{bmatrix}. \tag{4.3}$$

4.2 Row and Column Matrices

A row matrix has a single row of elements, and a column matrix has a single column of elements. In Equation 4.4 below, **A** is a row matrix, and **B** is a column matrix. In some situations, row and column matrices represent vectors. For example, if vectors $\mathbf{a} = (a_x, a_y, a_z)$ and $\mathbf{b} = (b_x, b_y\ b_z)$, then in matrix form we write them as

$$\mathbf{A} = \begin{bmatrix} a_x & a_y & a_z \end{bmatrix}, \quad \mathbf{B} = \begin{bmatrix} b_x \\ b_y \\ b_z \end{bmatrix}. \tag{4.4}$$

Of course, either **a** or **b** can be written as a row or a column matrix. The choice usually depends on the details of the matrix equation in which they might appear. More on this later.

Or we can represent a vector as a matrix product as follows:

$$\mathbf{P}=\begin{bmatrix} p_x & p_y & p_z \end{bmatrix}\begin{bmatrix} \mathbf{i} \\ \mathbf{j} \\ \mathbf{k} \end{bmatrix}. \tag{4.5}$$

This product is a 1×3 matrix premultiplying a 3×1 matrix, producing a 1×1 matrix, that is, a matrix with a single element (see Section 4.5 for much more on matrix multiplication):

$$\mathbf{P}=[p_x\mathbf{i}+p_y\mathbf{j}+p_z\mathbf{k}]. \tag{4.6}$$

4.3 Transpose of a Matrix

By interchanging the rows and columns of a matrix **A**, we obtain its transpose, $\mathbf{A}^{\mathrm{T}}$, so that $a_{ij}^{\mathrm{T}} = a_{ji}$, where the a_{ij}^{T} are the elements of the transpose of **A**. For example, if

$$\mathbf{A}=\begin{bmatrix} a & c & e \\ b & d & f \end{bmatrix}, \tag{4.7}$$

then

$$\mathbf{A}^{\mathrm{T}}=\begin{bmatrix} a & b \\ c & d \\ e & f \end{bmatrix}. \tag{4.8}$$

The transpose operator transforms a row matrix into a column matrix, and vice versa. So that if

$$\mathbf{A}=\begin{bmatrix} a_{11} & a_{12} & a_{13} \end{bmatrix} \tag{4.9}$$

and

$$\mathbf{B} = \mathbf{A}^{\mathrm{T}}, \tag{4.10}$$

then

$$b_{11}=a_{11},\; b_{21}=a_{12},\; b_{31}=a_{13}, \tag{4.11}$$

so that

$$\mathbf{B}=\begin{bmatrix} b_{11} \\ b_{21} \\ b_{31} \end{bmatrix}. \tag{4.12}$$

Note that for row and column matrices only a single subscript is necessary. In the examples above, a double subscript is used to indicate how the elements came to be.

4.4 Addition and Subtraction

Adding two matrices **A** and **B** produces a third matrix, **C**, whose elements are equal to the sum of the corresponding elements of **A** and **B**. Two matrices can be added or subtracted only if they are the same order. The resulting matrix **C** is also of the same order. For example,

$$\mathrm{A} + \mathrm{B} = \mathrm{C}, \tag{4.13}$$

or

$$a_{ij} + b_{ij} = c_{ij}. \tag{4.14}$$

Similarly, the difference of two matrices is another matrix, so that

$$\mathrm{A} - \mathrm{B} = \mathrm{D}, \tag{4.15}$$

or

$$a_{ij} + b_{ij} = d_{ij}. \tag{4.16}$$

We can add or subtract a sequence of more than two matrices as long as they are all the same order:

$$\mathrm{A} + \mathrm{B} - \mathrm{C} = \mathrm{D}, \tag{4.17}$$

or

$$a_{ij} + b_{ij} - c_{ij} = d_{ij} \tag{4.18}$$

Given two 3×2 matrices **A** and **B**, for example, where

$$\mathrm{A} = \begin{bmatrix} 2 & -3 \\ 1 & 4 \\ 0 & -1 \end{bmatrix} \tag{4.19}$$

and

$$\mathrm{B} = \begin{bmatrix} 5 & -1 \\ 2 & 9 \\ 2 & 7 \end{bmatrix}, \tag{4.20}$$

then

$$\begin{aligned} \mathrm{A} + \mathrm{B} &= \mathrm{C} \\ &= \begin{bmatrix} 2+5 & -3+(-1) \\ 1+2 & 4+9 \\ 0+2 & -1+7 \end{bmatrix} \\ &= \begin{bmatrix} 7 & -4 \\ 3 & 13 \\ 2 & 6 \end{bmatrix} \end{aligned} \tag{4.21}$$

and

$$\begin{aligned} \mathrm{A}-\mathrm{B} &= \mathrm{C} \\ &= \begin{bmatrix} 2-5 & -3+1 \\ 1-2 & 4-9 \\ 0-2 & -1-7 \end{bmatrix} \\ &= \begin{bmatrix} -3 & -2 \\ -1 & -5 \\ -2 & -8 \end{bmatrix}. \end{aligned} \tag{4.22}$$

Finally, note that matrix addition is commutative, which simply means that **A** + **B** = **B** + **A**.

4.5 Multiplication

Multiplying a matrix **A** by a constant k produces a new matrix **B** of the same order. Each element of **B** is obtained by multiplying the corresponding element of **A** by the constant k, so that

$$\mathbf{B} = k\mathbf{A}, \tag{4.23}$$

or

$$b_{ij} = ka_{ij}. \tag{4.24}$$

The product **AB** of two matrices is another matrix, **C**. This operation is possible if and only if the number of columns of the first matrix is equal to the number of rows of the second matrix. If **A** is $m \times n$ and **B** is $n \times p$, then **C** is $m \times p$. When this condition is satisfied, we say that the matrices **A** and **B** are conformable for multiplication, and we write this as **C** = **AB**. We describe this product by stating that **A** premultiplies **B**, or **B** postmultiplies **A**.

The product of two matrices is not commutative unless one of them is the identity matrix (see Section 5.1). Thus, in general,

$$\mathbf{AB} \neq \mathbf{BA}. \tag{4.25}$$

The product of two matrices in terms of their elements is

$$c_{ij} = \sum_{k=1}^{n} a_{ik} b_{kj}, \tag{4.26}$$

where

$$c_{ij} = a_{i1}b_{1j} + a_{i2}b_{2j} + \ldots + a_{in}b_{nj}, \tag{4.27}$$

and n is equal to the number of columns of **A** and the number of rows of **B**.

This somewhat unusual expression makes more sense intuitively with an example: Given the 3×3 matrix **A** and the 3×2 matrix **B**, they are conformable, and their product is the 3×2 matrix **C**. To see this more graphically, we arrange the matrices as follows (Figure 4.1):

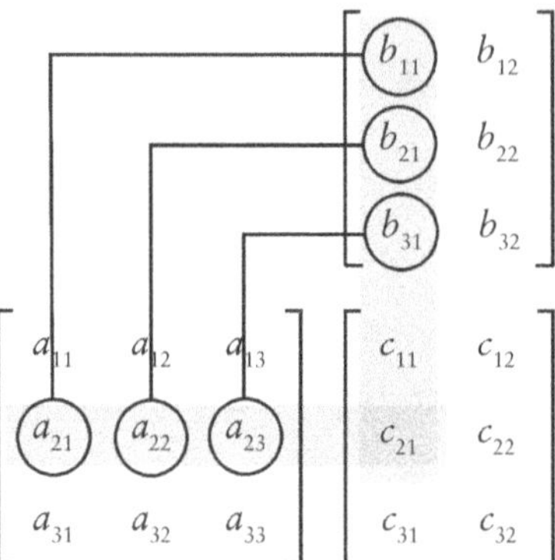

Figure 4.1 The product of two matrices

Each element c_{ij} is determined by summing the pairwise products of elements in row i of **A** with corresponding elements in column j of **B**. For example,

$$c_{21} = a_{21}b_{11} + a_{22}b_{21} + a_{23}b_{31} \tag{4.28}$$

This also illustrates why **A** and **B** must be conformable and how their orders determine the order of **C**.

Here is one more example. Let

$$\mathrm{A} = \begin{bmatrix} 2 & -1 \\ 3 & 4 \end{bmatrix} \tag{4.29}$$

and

$$\mathrm{B} = \begin{bmatrix} 1 & 0 & 3 \\ 5 & 2 & 4 \end{bmatrix}. \tag{4.30}$$

Then

$$\begin{aligned} \mathrm{C} &= \begin{bmatrix} 2\times1-1\times5 & 2\times0-1\times2 & 2\times3-1\times4 \\ 3\times1+4\times5 & 3\times0+4\times2 & 3\times3+4\times4 \end{bmatrix} \\ &= \begin{bmatrix} -3 & -2 & 2 \\ 23 & 8 & 25 \end{bmatrix}. \end{aligned} \tag{4.31}$$

This looks like computing the dot or inner product of rows and columns: that is, between the rows of **A** and the columns of **B**. Clearly we are treating the rows and columns as if they are vectors. This outcome proves to be very useful later in our studies.

Furthermore, it is possible to construct matrices **A** and **B** such that **A** ≠ 0, **B** ≠ 0, but **AB** = 0. A zero or null matrix is one for which all elements are zero.

And here are even more special characteristics of matrix multiplication:

$$(\mathbf{AB})^{\mathrm{T}} = \mathbf{B}^{\mathrm{T}}\mathbf{A}^{\mathrm{T}} \tag{4.32}$$

and

$$(\mathbf{AB})\mathbf{C} = \mathbf{A}(\mathbf{BC}). \tag{4.33}$$

A simple example will demonstrate Equation 4.32. Let

$$\mathbf{A} = \begin{bmatrix} a & b \\ c & d \end{bmatrix}, \tag{4.34}$$

$$\mathbf{B} = \begin{bmatrix} e & f \\ g & h \end{bmatrix}, \tag{4.35}$$

and

$$\mathbf{C} = \begin{bmatrix} i & j \\ k & l \end{bmatrix}. \tag{4.36}$$

Then

$$\mathbf{AB} = \begin{bmatrix} ae+bg & af+bh \\ ce+dg & cf+dh \end{bmatrix}, \tag{4.37}$$

$$\mathbf{A}^{\mathrm{T}} = \begin{bmatrix} a & c \\ b & d \end{bmatrix}, \tag{4.38}$$

and

$$\mathbf{B}^{\mathrm{T}} = \begin{bmatrix} e & g \\ f & h \end{bmatrix}, \tag{4.39}$$

so that

$$(\mathbf{AB})^{\mathrm{T}} = \begin{bmatrix} ae+bg & ce+bg \\ af+bh & cf+dh \end{bmatrix} \tag{4.40}$$

and finally

$$\mathbf{B}^{\mathrm{T}}\mathbf{A}^{\mathrm{T}} = \begin{bmatrix} ae+bg & ce+dg \\ af+bh & cf+dh \end{bmatrix}. \tag{4.41}$$

We see that these last two equations are equal, which demonstrates the assertion of Equation 4.32.

Next, let's check Equation 4.33, and the associative character of matrix multiplication. First, from Equation 4.37 and Equation 4.36, earlier,

$$\mathbf{AB} = \begin{bmatrix} ae+bg & af+bh \\ ce+dg & cf+dh \end{bmatrix}$$
$$\mathbf{C} = \begin{bmatrix} i & j \\ k & l \end{bmatrix}. \tag{4.42}$$

Then

$$\mathbf{BC} = \begin{bmatrix} (ei+fk) & (ej+fl) \\ (gi+hk) & (gj+hl) \end{bmatrix}, \tag{4.43}$$

so that

$$\begin{aligned}(\mathbf{AB})\mathbf{C} &= \begin{bmatrix} (ae+bg)i+(af+bh)k & (ae+bg)j+(af+bh)l \\ (ce+dg)i+(ef+dh)k & (ce+dg)j+(cf+dh)l \end{bmatrix} \\ &= \begin{bmatrix} (ae+bg)i+(af+bh)k & (ae+bg)j+(af+bh)l \\ (ce+dg)i+(cf+dh)k & (ce+dg)j+(cf+dh)l \end{bmatrix}\end{aligned} \tag{4.44}$$

and

$$\mathbf{A}(\mathbf{BC}) = \begin{bmatrix} a(ei+fk)+b(gi+hk) & a(ej+fl)+b(gj+hl) \\ c(ei+fk)+d(gi+hk) & c(ej+fl)+d(gj+hl) \end{bmatrix}. \tag{4.45}$$

Again, we see that these last two equations are equal, which demonstrates the assertion of Equation 4.33.

4.6 Partitioning a Matrix

It is sometimes necessary to partition a matrix into submatrices and to treat it as a matrix whose elements are themselves matrices. We do this to simplify and speed up computations involving large sparse matrices (see below). So let's partition the matrix **M**:

$$\mathbf{M} = \left[\begin{array}{c|c} \mathbf{M}_{11} & \mathbf{M}_{12} \\ \hline \mathbf{M}_{21} & \mathbf{M}_{22} \end{array}\right], \tag{4.46}$$

where $\mathbf{M}_{11}$ and $\mathbf{M}_{12}$ must necessarily have the same number of rows (and similarly for $\mathbf{M}_{21}$ and $\mathbf{M}_{22}$), and $\mathbf{M}_{11}$ and $\mathbf{M}_{21}$ must necessarily have the same number of columns (and similarly for $\mathbf{M}_{12}$ and $\mathbf{M}_{22}$). For example, if

$$\mathbf{A} = \left[\begin{array}{c|c|c} \mathbf{A}_{11} & \mathbf{A}_{12} & \mathbf{A}_{13} \\ \hline \mathbf{A}_{21} & \mathbf{A}_{22} & \mathbf{A}_{23} \end{array}\right], \quad \mathbf{B} = \left[\begin{array}{c|c|c} \mathbf{B}_{11} & \mathbf{B}_{12} & \mathbf{B}_{13} \\ \hline \mathbf{B}_{21} & \mathbf{B}_{22} & \mathbf{B}_{23} \end{array}\right], \tag{4.47}$$

and

$$\mathbf{C} = \mathbf{A} + \mathbf{B}, \tag{4.48}$$

then

$$\mathbf{C} = \left[\begin{array}{c|c|c} \mathbf{A}_{11}+\mathbf{B}_{11} & \mathbf{A}_{12}+\mathbf{B}_{12} & \mathbf{A}_{13}+\mathbf{B}_{13} \\ \hline \mathbf{A}_{21}+\mathbf{B}_{21} & \mathbf{A}_{22}+\mathbf{B}_{22} & \mathbf{A}_{23}+\mathbf{B}_{23} \end{array}\right], \tag{4.49}$$

where the $\mathbf{A}_{ij}$ and $\mathbf{B}_{ij}$ are conformable.

Here is an example of the multiplication of partitioned matrices. The matrices must be conformable for multiplication before and after partitioning, as in the following example:

$$\mathbf{A} = \left[\begin{array}{cc|cc|c} a_{11} & a_{12} & a_{13} & a_{14} & a_{15} \\ a_{21} & a_{22} & a_{23} & a_{24} & a_{25} \\ \hline a_{31} & a_{32} & a_{33} & a_{34} & a_{35} \\ a_{41} & a_{42} & a_{43} & a_{44} & a_{45} \end{array}\right] = \begin{bmatrix} \mathbf{A}_{11} & \mathbf{A}_{12} & \mathbf{A}_{13} \\ \mathbf{A}_{21} & \mathbf{A}_{22} & \mathbf{A}_{23} \end{bmatrix}, \tag{4.50}$$

$$\mathbf{B} = \left[\begin{array}{cc|c} b_{11} & b_{12} & b_{13} \\ b_{21} & b_{22} & b_{23} \\ \hline b_{31} & b_{32} & b_{33} \\ b_{41} & b_{42} & b_{43} \\ \hline b_{51} & b_{52} & b_{53} \end{array}\right] = \begin{bmatrix} \mathbf{B}_{11} & \mathbf{B}_{12} \\ \mathbf{B}_{21} & \mathbf{B}_{22} \\ \mathbf{B}_{31} & \mathbf{B}_{32} \end{bmatrix}, \tag{4.51}$$

$$\mathbf{AB} = \begin{bmatrix} \mathbf{A}_{11}\mathbf{B}_{11}+\mathbf{A}_{12}\mathbf{B}_{21}+\mathbf{A}_{13}\mathbf{B}_{31} & \mathbf{A}_{11}\mathbf{B}_{12}+\mathbf{A}_{12}\mathbf{B}_{22}+\mathbf{A}_{13}\mathbf{B}_{32} \\ \mathbf{A}_{21}\mathbf{B}_{11}+\mathbf{A}_{22}\mathbf{B}_{21}+\mathbf{A}_{23}\mathbf{B}_{31} & \mathbf{A}_{21}\mathbf{B}_{12}+\mathbf{A}_{22}\mathbf{B}_{22}+\mathbf{A}_{23}\mathbf{B}_{32} \end{bmatrix}. \tag{4.52}$$

Determining the products $A_{11}B_{11}$, $A_{12}B_{21}$, etc., and performing the indicated sums completes this computation.

So what is a sparse matrix? Most of the elements of a sparse matrix are zero, and that is its distinguishing characteristic. Conversely, if most of the elements are nonzero, the matrix is dense. There are numerical methods that take advantage of a matrix's sparseness to decrease the number of matrix computations and to increase the speed of matrix computations.

4.7 Summary of Matrix Properties

To reinforce your understanding of these properties, create some example matrices and demonstrate the truth of the following:

1. $\mathbf{A} + \mathbf{B} = \mathbf{B} + \mathbf{A}$
2. $\mathbf{A} + (\mathbf{B} + \mathbf{C}) = (\mathbf{A} + \mathbf{B}) + \mathbf{C}$
3. $k(\mathbf{A} + \mathbf{B}) = k\mathbf{A} + k\mathbf{B}$
4. $(k + l)\mathbf{A} = k\mathbf{A} + l\mathbf{A}$
5. $k(l\mathbf{A}) = (kl)\,\mathbf{A} = l(k\mathbf{A})$
6. $(\mathbf{AB})\mathbf{C} = \mathbf{A}(\mathbf{BC})$
7. $\mathbf{A}(\mathbf{B} + \mathbf{C}) = \mathbf{AB} + \mathbf{AC}$
8. $(\mathbf{A} + \mathbf{B})\mathbf{C} = \mathbf{AC} + \mathbf{BC}$
9. $\mathbf{A}(k\mathbf{B}) = k\mathbf{AB} = (k\mathbf{A})\mathbf{B}$
10. $(\mathbf{A} + \mathbf{B})^T = \mathbf{A}^T + \mathbf{B}^T$
11. $(k\mathbf{A})^T = k\mathbf{A}^T$
12. $(\mathbf{AB})^T = \mathbf{B}^T\mathbf{A}^T$
13. If $\mathbf{AA}^{-1} = \mathbf{I}$ and $\mathbf{A}^{-1}\mathbf{A} = \mathbf{I}$, then **A** is nonsingular.
14. If $|\mathbf{A} = 0|$, then **A** is singular.
15. If $\mathbf{AA}^T = \mathbf{I}$, then **A** is orthogonal.

4.8 Exercises

4.1. Given $\mathbf{A} = \begin{bmatrix} 7 & 3 & -1 \\ 2 & -5 & 6 \end{bmatrix}$ and $\mathbf{B} = \begin{bmatrix} 1 & 5 & 6 \\ -4 & -2 & 3 \end{bmatrix}$,

a. Find $\mathbf{A} + \mathbf{B}$.

b. Find $\mathbf{A} - \mathbf{B}$.

4.2. Given $\mathbf{A} = \begin{bmatrix} 3 & 7 & -2 \end{bmatrix}$, find $-\mathbf{A}$.

4.3. Given $\mathbf{A}=\begin{bmatrix} 1 & 5 & 2 \\ 0 & -1 & 4 \end{bmatrix}$, $\mathbf{B}=\begin{bmatrix} 6 & 1 & 3 \\ 0 & 9 & 2 \end{bmatrix}$, and $\mathbf{C}=\begin{bmatrix} 4 & 1 & 1 \\ 5 & 8 & 3 \end{bmatrix}$,

a. Find $\mathbf{A} + 2\mathbf{A}$.

b. Find $\mathbf{B} + \mathbf{B}$.

c. Find $2\mathbf{A} + \mathbf{B}$.

d. Find $\mathbf{A} - \mathbf{B} + \mathbf{C}$.

e. Find $\mathbf{A} - 2\mathbf{B} - \mathbf{C}$.

4.4. Given $\mathbf{A}=\begin{bmatrix} 1 & -4 \\ 3 & 0 \end{bmatrix}$, find $1.5\mathbf{A}$.

4.5. Find $\mathbf{I}^T$.

4.6. Given $\mathbf{A}=\begin{bmatrix} 5 & 3 & 8 \\ -1 & 4 & 7 \\ 0 & 1 & 1 \end{bmatrix}$ and $\mathbf{B}=\begin{bmatrix} 6 & 7 \\ 10 & 9 \\ 2 & -3 \end{bmatrix}$, find $\mathbf{AB}$.

4.7. Given $\mathbf{A}=\begin{bmatrix} 2 & 1 \\ 3 & 4 \end{bmatrix}$ and $\mathbf{B}=\begin{bmatrix} 6 \\ 3 \end{bmatrix}$, find $\mathbf{AB}$.

4.8. Given $\mathbf{A}=\begin{bmatrix} 4 & 0 & 7 \\ 5 & 1 & 2 \end{bmatrix}$, find $\mathbf{A}^T$.

4.9. Given $\mathbf{A}=\begin{bmatrix} 4 & -2 \\ 1 & 0 \\ 6 & 7 \end{bmatrix}$ and $\mathbf{B}=\begin{bmatrix} 1 & 1 \\ 5 & 2 \\ 2 & 4 \end{bmatrix}$,

a. Find $(\mathbf{A}^T)^T$.

b. Find $(\mathbf{A} + \mathbf{B})^T$.

c. Find $\mathbf{A}^T + \mathbf{B}^T$.

d. Find $\mathbf{B}^T + \mathbf{A}^T$.

4.10. Find the product $\begin{bmatrix} t^2 & t & 1 \end{bmatrix}\begin{bmatrix} a_x & a_y \\ b_x & b_y \\ c_x & c_y \end{bmatrix}$.

4.11. Given $\mathbf{A} = \begin{bmatrix} 1 & 0 & 0 \\ 0 & 1 & 0 \\ 0 & 0 & 1 \end{bmatrix}$, $\mathbf{B} = \begin{bmatrix} 7 \\ 4 \\ 9 \\ 5 \end{bmatrix}$, and $\mathbf{C} = [\ 1 \quad -2 \quad 4 \quad 6\ ]$,

a. Find a_{23}.

b. Find a_{32}.

c. Find b_{31}.

d. Find c_{14}.

e. What is the order of **A**?

f. What is the order of **B**?

g. What is the order of **C**?

h. Which is the column matrix?

i. Which is the row matrix?

j. Which is the identity matrix?

4.12. If **P** = **ABC** and the order of **A** is 1×4, the order of **B** is 4×4, and the order of **C** is 4×3, then what is the order of **P**?

5 Special Matrices

This chapter looks at special matrices. They all happen to be square matrices. A square matrix has the same number of rows and columns. Some are the gateway to geometric transformations, eigenvectors and tensors, and more.

5.1 Identity Matrix

An important matrix in matrix algebra is the identity matrix. It has elements only on the main diagonal, and they are all equal to 1. It is denoted by the symbol **I**. The identity matrix is necessarily both a square matrix and a diagonal matrix (Section 5.2). The 3×3 identity matrix is

$$\mathbf{I}=\begin{bmatrix} 1 & 0 & 0 \\ 0 & 1 & 0 \\ 0 & 0 & 1 \end{bmatrix}. \tag{5.1}$$

The notation δ_{ij}, known as the Kronecker delta, represents the elements of **I**, where

$$\begin{aligned} \delta_{ij} &= 0 \text{ if } i \neq j, \\ \delta_{ij} &= 1 \text{ if } i = j. \end{aligned} \tag{5.2}$$

Here is another characteristic of the identity matrix: Multiplying any matrix **A** of order $m \times n$ by the identity matrix reproduces the original matrix: **AI** = **A** or **IA** = **A**, where **I** must be order $n \times n$ if it postmultiplies **A**, and have order $m \times m$ if it premultiplies **A**. For example,

$$\begin{bmatrix} a_{11} & a_{12} & \cdots & a_{1n} \\ a_{21} & a_{22} & \cdots & a_{2n} \\ \vdots & \vdots & \ddots & \vdots \\ a_{m1} & a_{m2} & \cdots & a_{mn} \end{bmatrix}\begin{bmatrix} 1 & 0 & \cdots & 0 \\ 0 & 1 & \cdots & 0 \\ \vdots & \vdots & \ddots & \vdots \\ 0 & 0 & \cdots & 1 \end{bmatrix}=\begin{bmatrix} a_{11} & a_{12} & \cdots & a_{1n} \\ a_{21} & a_{22} & \cdots & a_{2n} \\ \vdots & \vdots & \ddots & \vdots \\ a_{m1} & a_{m2} & \cdots & a_{mn} \end{bmatrix}. \tag{5.3}$$

We see that multiplication by the identity matrix is commutative, so that

$$\mathbf{AI} = \mathbf{IA}. \tag{5.4}$$

5.2 Diagonal Matrix

A square matrix that has zero elements everywhere except on the main diagonal is a diagonal matrix. (The main diagonal runs from the upper-left-corner element to the lower-right-corner element.) This means that

$$a_{ij} = 0 \text{ if } i \neq j, \tag{5.5}$$

so that

$$\mathrm{A} = \begin{bmatrix} a_{11} & 0 & 0 & 0 \\ 0 & a_{22} & 0 & 0 \\ 0 & 0 & a_{33} & 0 \\ 0 & 0 & 0 & a_{44} \end{bmatrix}. \tag{5.6}$$

The determinant (Section 5.9) of a diagonal matrix is simply the product of the diagonal elements. For example, the determinant of the matrix above is

$$|\mathrm{A}| = a_{11}a_{22}a_{33}a_{44}. \tag{5.7}$$

The product of two diagonal matrices is another diagonal matrix. Here is an example: Given

$$\mathrm{A} = \begin{bmatrix} a_{11} & 0 & 0 \\ 0 & a_{22} & 0 \\ 0 & 0 & a_{33} \end{bmatrix} \quad \text{and} \quad \mathrm{B} = \begin{bmatrix} b_{11} & 0 & 0 \\ 0 & b_{22} & 0 \\ 0 & 0 & b_{33} \end{bmatrix}, \tag{5.8}$$

then

$$\mathrm{AB} = \begin{bmatrix} a_{11}b_{11} & 0 & 0 \\ 0 & a_{22}b_{22} & 0 \\ 0 & 0 & a_{33}b_{33} \end{bmatrix}. \tag{5.9}$$

If all the a_{ij} are equal, then the diagonal matrix is a scalar matrix. For example,

$$\mathrm{A} = \begin{bmatrix} 2 & 0 & 0 \\ 0 & 2 & 0 \\ 0 & 0 & 2 \end{bmatrix}, \tag{5.10}$$

or we can write this as

$$\mathrm{A} = 2\begin{bmatrix} 1 & 0 & 0 \\ 0 & 1 & 0 \\ 0 & 0 & 1 \end{bmatrix} = 2\mathbf{I}\cdot \tag{5.11}$$

The general form of a scalar matrix is simply

$$\mathbf{A} = a\begin{bmatrix} 1 & 0 & 0 \\ 0 & 1 & 0 \\ 0 & 0 & 1 \end{bmatrix} = a\mathbf{I}. \tag{5.12}$$

A scalar matrix is a scalar multiplier of a vector, changing its magnitude but not its direction. For example, if

$$\mathbf{A} = \begin{bmatrix} k & 0 & 0 \\ 0 & k & 0 \\ 0 & 0 & k \end{bmatrix} \text{ and } \mathbf{P} = \begin{bmatrix} p_x \\ p_y \\ p_z \end{bmatrix}, \tag{5.13}$$

where **P** is the matrix of components of the vector **p**, then

$$\mathbf{AP} = \begin{bmatrix} kp_x \\ kp_y \\ kp_z \end{bmatrix}. \tag{5.14}$$

5.3 Trace of a Matrix

The trace, tr(**A**), of a square matrix **A** is the sum of its diagonal entries:

$$\mathrm{tr}(\mathbf{A}) = a_{11} + a_{22} + \ldots + a_{nn}, \tag{5.15}$$

or

$$\mathrm{tr}(\mathbf{A}) = \sum_{i=1}^{n} a_{ii}, \tag{5.16}$$

where a_{nn} denotes the entry on the nth row and nth column of **A**. Note that the trace is only defined for square matrices. We will soon see that it plays an important role in expressing the characteristic equation associated with eigenvectors and eigenvalues.

Although matrix multiplication is not commutative, as we have seen, the trace of the product of two matrices is independent of the order of the matrices, so that

$$\mathrm{tr}(\mathbf{AB}) = \mathrm{tr}(\mathbf{BA}). \tag{5.17}$$

This is apparent from the tensor definition of matrix multiplication:

$$\mathrm{tr}(\mathbf{AB}) = \mathrm{tr}(\mathbf{BA}) = a_{ij}b_{ji}. \tag{5.18}$$

Also, the trace of a matrix is equal to that of its transpose; that is,

$$\mathrm{tr}(\mathbf{A}) = \mathrm{tr}(\mathbf{A}^{\mathrm{T}}). \tag{5.19}$$

For example, if

$$\mathbf{A} = \begin{bmatrix} a & b & c \\ d & e & f \\ g & h & i \end{bmatrix}, \tag{5.20}$$

then

$$\mathbf{A}^{\mathrm{T}} = \begin{bmatrix} a & d & g \\ b & e & h \\ c & f & i \end{bmatrix}, \tag{5.21}$$

and we see that the diagonal elements of **A** and $\mathbf{A}^{\mathrm{T}}$ are equal, so their traces are equal.

5.4 Symmetric Matrix

A matrix whose elements are symmetric about the main diagonal is a symmetric matrix. If **A** is a symmetric matrix, then $a_{ij} = a_{ji}$; for example,

$$\mathbf{A} = \begin{bmatrix} 3 & 0 & -4 & 1 \\ 0 & 6 & 2 & 3 \\ -4 & -2 & 5 & 7 \\ 1 & 3 & 7 & -2 \end{bmatrix}. \tag{5.22}$$

A matrix is antisymmetric (or skew symmetric) if $a_{ij} = -a_{ji}$; for example,

$$\mathbf{A} = \begin{bmatrix} 3 & 0 & -4 & 1 \\ 0 & 6 & 2 & 3 \\ 4 & -2 & 5 & 7 \\ -1 & -3 & -7 & -2 \end{bmatrix}. \tag{5.23}$$

5.5 Orthogonal Matrix

Rotation and reflection transformation matrices are orthogonal. Conversely, orthogonal matrices imply orthogonal transformations (Chapter 9). An orthogonal transformation preserves the lengths of vectors and the angles between them.

Given a real, square matrix **A** of order n, if

$$\mathbf{A}\mathbf{A}^{\mathrm{T}} = \mathbf{A}^{\mathrm{T}}\mathbf{A} = \mathbf{I}, \tag{5.24}$$

which we can also write as

$$\sum_{k=1}^{n} a_{ik}a_{jk} = \delta_{ij}, \tag{5.25}$$

then

$$A^T = A^{-1}, \tag{5.26}$$

and **A** is an orthogonal matrix.

If **A** is orthogonal, then

$$|A| = 1, \tag{5.27}$$

which requires computing the determinant of **A** (see Section 5.9).

5.6 Inverse of a Matrix

Matrix algebra does not define a division operation, but it does include a process for finding the inverse of a matrix. The inverse of a square matrix **A** is A^{-1}. A matrix and its inverse satisfy

$$AA^{-1} = A^{-1}A = I. \tag{5.28}$$

The elements of A^{-1} are

$$a_{ij}^{-1} = \frac{(-1)^{i+j}\left|A'_{ji}\right|}{|A|}, \tag{5.29}$$

where $\left|A'_{ji}\right|$ denotes the determinant of the $(n - 1) \times (n - 1)$ matrix derived from **A** by deleting row j and column i from **A** (note the subscript order on A'_{ji}). If A^{-1} exists, then $|A| \neq 0$. Again, refer to Section 5.9 for a refresher on determinants.

For example, find A^{-1} when

$$A = \begin{bmatrix} 2 & -1 & 3 \\ 1 & 6 & -4 \\ 5 & 0 & 8 \end{bmatrix}. \tag{5.30}$$

First, compute $|A| = 34$. Then replace each element of **A** with its cofactor and transpose the result to obtain

$$C^T = \begin{bmatrix} 48 & 8 & -14 \\ -28 & 1 & 11 \\ -30 & -5 & 13 \end{bmatrix}. \tag{5.31}$$

We use $\mathbf{A}^{-1} = \mathbf{C}^{T}/|\mathbf{A}|$ to obtain

$$\mathrm{A}^{-1} = \frac{1}{34}\begin{bmatrix} 48 & 8 & -14 \\ -28 & 1 & 11 \\ -30 & -5 & 13 \end{bmatrix}. \tag{5.32}$$

Note that for $\mathbf{A}^{-1}$ to exist at all, $|\mathbf{A}| \neq 0$. It is easy to check the correctness of $\mathbf{A}^{-1}$ because $\mathbf{AA}^{-1} = \mathbf{A}^{-1}\mathbf{A} = \mathbf{I}$.

Matrix inversion is comparable to taking the reciprocal of an expression in the course of solving an algebraic equation. For the expression

$$3x = 8, \tag{5.33}$$

the value of x is computed by operating on 8 with the reciprocal of 3, so that

$$\left(\frac{1}{3}\right)3x = \left(\frac{1}{3}\right)8, \quad \text{or} \quad x = \frac{8}{3} \tag{5.34}$$

A similar procedure applies to matrix algebra, using matrix inversion. For example, let's examine the following set of n simultaneous linear equations with n unknowns:

$$\begin{aligned} a_{11}x_1 + a_{12}x_2 + \ldots + a_{1n}x_n &= b_1, \\ a_{21}x_1 + a_{22}x_2 + \ldots + a_{2n}x_n &= b_2, \\ \vdots \qquad\qquad & \quad \vdots \\ a_{n1}x_1 + a_{n2}x_2 + \ldots + a_{nn}x_n &= b_n. \end{aligned} \tag{5.35}$$

We write this set of equations as the following matrix equation:

$$\mathbf{AX} = \mathbf{B}. \tag{5.36}$$

Premultiply both sides by $\mathbf{A}^{-1}$ to obtain

$$\mathbf{A}^{-1}\,\mathbf{AX} = \mathbf{A}^{-1}\,\mathbf{B}. \tag{5.37}$$

Because $\mathbf{A}^{-1}\,\mathbf{A} = \mathbf{I}$ and $\mathbf{IX} = \mathbf{X}$, Equation 5.32 simplifies to

$$\mathbf{X} = \mathbf{A}^{-1}\,\mathbf{B}. \tag{5.38}$$

5.7 Collinear Vectors, Eigenvalues, and Eigenvectors

Certain forms of matrix multiplication can change the magnitude or direction of a vector. For example, if $\mathbf{P}$ is an $n \times 1$ column matrix representing an n-component vector, then multiplying it by an $n \times n$ transformation matrix $\mathbf{A}$ produces

$$\mathbf{P}' = \mathbf{AP}, \tag{5.39}$$

where $\mathbf{P}'$ is the matrix containing the new, or transformed, vector.

It is possible to find a scalar constant λ (lowercase Greek lambda) such that

$$\mathbf{P}' = \lambda\mathbf{P}, \tag{5.40}$$

or

$$\mathbf{AP} = \lambda\mathbf{P}. \tag{5.41}$$

Every vector for which this is true for a given transformation matrix **A** is an eigenvector of **A**, and λ is the eigenvalue of **A** corresponding to the vector **P**. Matrix **A** transforms its eigenvector(s) into a collinear vector. The corresponding eigenvalue(s) is equal to the ratio of the magnitudes of the two collinear vectors. (*Eigenvalue* is from the German *eigenwerte*, meaning proper value.)

Here is a geometric interpretation of an eigenvector: Figure 5.1a shows what to expect for the general case. Multiplying any vector **Q** by an arbitrary but conformable matrix **A** produces some new vector $\mathbf{Q}'$. However, there are certain vectors **P** that when multiplied by **A** lie on the line *OP*; $\mathbf{P}'$ is an example of this (Figure 5.1b and 5.1c).

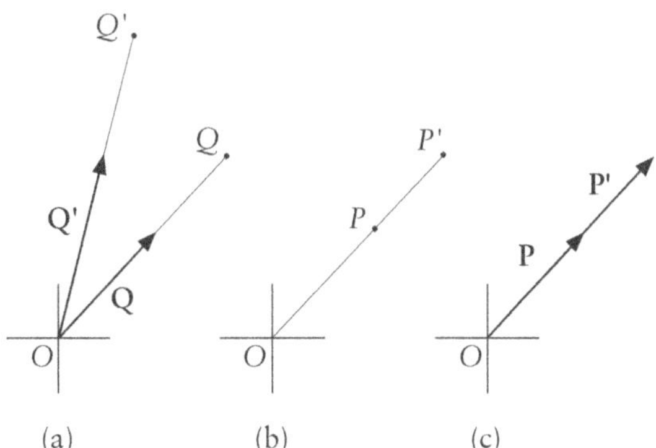

Figure 5.1 A geometric interpretation of eigenvectors

Rewrite $\mathbf{AP} = \lambda\mathbf{P}$ as $(\mathbf{A} - \lambda\mathbf{I})\mathbf{P} = 0$, which has nontrivial solutions if $\mathbf{P} \neq 0$, so that $|\mathbf{A} - \lambda\mathbf{I}| = 0$. This is what mathematicians call the characteristic equation, and its solutions are eigenvalues of **A**. Then the solutions of $(\mathbf{A} - \lambda_i\mathbf{I})\mathbf{P} = 0$ are the eigenvectors corresponding to λ_i.

Here is an example for a 2x2 matrix. Let

$$\mathbf{A} = \begin{bmatrix} 5 & -1 \\ 3 & 1 \end{bmatrix}. \tag{5.42}$$

Then

$$\begin{vmatrix} 5-\lambda & -1 \\ 3 & 1-\lambda \end{vmatrix} = 0. \tag{5.43}$$

Solving this determinant produces the characteristic equation $\lambda^2 - 6\lambda + 8 = 0$, whose solution yields the eigenvalues $\lambda_1 = 2$ and $\lambda_2 = 4$.

Again, using $\mathbf{AP} = \lambda\mathbf{P}$, we find

$$\begin{bmatrix} 5 & -1 \\ 3 & 1 \end{bmatrix}\begin{bmatrix} x \\ y \end{bmatrix} = \lambda\begin{bmatrix} x \\ y \end{bmatrix}. \tag{5.44}$$

Substitute the eigenvalue $\lambda_1 = 2$ to yield $5x - y = 2x$ or $3x = y$, and $3x + y = 2y$ or $3x = y$. This shows that a family of vectors is associated with the eigenvalue $\lambda_1 = 2$ that has the form

$$\begin{bmatrix} k \\ 3k \end{bmatrix}, \tag{5.45}$$

where $k \neq 0$. In other words, any vector whose y component is exactly three times as great as its x component is an eigenvector of **A**.

For the other eigenvalue, $\lambda_2 = 4$, we have $5x - y = 4x$ or $x = y$, and $3x + y = 4y$ or $x = y$, so that λ_2 has eigenvectors of the form

$$\begin{bmatrix} k \\ k \end{bmatrix}. \tag{5.46}$$

Any vector whose x and y components are equal is an eigenvector of **A**.

For the general case of a 2×2 matrix, if

$$\mathrm{A} = \begin{bmatrix} a & b \\ c & d \end{bmatrix}, \tag{5.47}$$

then we compute the eigenvalues from

$$\begin{vmatrix} a-\lambda & b \\ c & d-\lambda \end{vmatrix} = 0, \tag{5.48}$$

or

$$\lambda^2 - (a+d)\lambda + (ad - bc) = 0. \tag{5.49}$$

There are, of course, two roots to this equation, the two eigenvalues. These may be real and distinct, equal or complex. The eigenvalues are used to compute values of the corresponding eigenvectors, which reveal two possible forms for the ratio of the vector components, depending on the value of λ; thus,

$$\frac{p_x}{p_y} = \frac{-b}{a - \lambda_i} \tag{5.50}$$

or

$$\frac{p_x}{p_y} = \frac{-(d - \lambda_i)}{c}, \tag{5.51}$$

where $i = 1, 2$. We can generalize this process to apply to $n \times n$ matrices, but this is beyond the scope of our studies here.

As we have seen above, an eigenvector $\mathbf{P}_i$ is associated with each eigenvalue λ_i of a square matrix $\mathbf{A}$, which we can now write as

$$\mathbf{AP}_i = \lambda_i \mathbf{P}_i. \tag{5.52}$$

This makes it possible to construct a square matrix $\mathbf{E}$ of order n whose columns are the eigenvectors $\mathbf{P}_i$ of $\mathbf{A}$. This allows us to rewrite $\mathbf{AP}_i = \lambda_i \mathbf{P}_i$ as

$$\mathbf{AE} = \mathbf{E}\Lambda, \tag{5.53}$$

where Λ is a diagonal matrix whose elements are the eigenvalues of $\mathbf{A}$:

$$\Lambda = \begin{bmatrix} \lambda_1 & 0 & \dots & 0 \\ 0 & \lambda_2 & \dots & 0 \\ \vdots & \vdots & \ddots & \vdots \\ 0 & 0 & \dots & \lambda_n \end{bmatrix}. \tag{5.54}$$

If the eigenvalues are distinct, then the matrix $\mathbf{E}$ is nonsingular, and premultiplying both sides of the equation by $\mathbf{E}^{-1}$ yields

$$\mathbf{E}^{-1}\mathbf{AE} = \Lambda. \tag{5.55}$$

This means that by using the matrix of eigenvectors and its inverse, any matrix $\mathbf{A}$ with distinct eigenvalues can be transformed into a diagonal matrix whose elements are the eigenvalues of $\mathbf{A}$. Mathematicians call this process the diagonalization of the matrix $\mathbf{A}$.

5.8 Similar Matrices

If a matrix $\mathbf{A}$ is premultiplied and postmultiplied by another matrix and its inverse, respectively, then the resulting matrix $\mathbf{B}$ is a similarity transformation of $\mathbf{A}$; thus,

$$\mathbf{B} = \mathbf{TAT}^{-1}, \tag{5.56}$$

and because of this relationship we say that $\mathbf{A}$ and $\mathbf{B}$ are similar matrices. Similar matrices have equal determinants, the same characteristic equation, and the same eigenvalues, although not necessarily the same eigenvectors. Similarity transformations preserve eigenvalues.

If **A** is similar to a diagonal matrix **D**, then

$$\mathbf{D} = \begin{bmatrix} \lambda_1 & 0 & \dots & 0 \\ 0 & \lambda_2 & \dots & 0 \\ \vdots & \vdots & \ddots & \vdots \\ 0 & 0 & \dots & \lambda_n \end{bmatrix}, \tag{5.57}$$

where $\lambda_1, \lambda_2, \ldots, \lambda_n$ are the eigenvalues of **A**. However, for this to be true, there must be a nonsingular matrix **S** such that $\mathbf{SAS}^{-1} = \mathbf{D}$. Obviously, the elements on the main diagonal of a diagonal matrix **D** are its eigenvalues. Therefore, the eigenvalues of **D** are the eigenvalues of **A**. Finally, if the eigenvalues of **A** are distinct, then it is similar to a diagonal matrix.

5.9 Determinants

The determinant of a square matrix is a single value obtained by combining the elements of the matrix in a specific way. Before getting into the process for doing this, let's look at two examples.

The determinant of a 2×2 matrix **A** is $|\mathbf{A}|$, where

$$\mathbf{A} = \begin{bmatrix} a_{11} & a_{12} \\ a_{21} & a_{22} \end{bmatrix} \tag{5.58}$$

and its determinant is

$$|\mathbf{A}| = a_{11}a_{22} - a_{12}a_{21}. \tag{5.59}$$

The determinant of a 3×3 matrix **A** is $|\mathbf{A}|$, where

$$\mathbf{A} = \begin{bmatrix} a_{11} & a_{12} & a_{13} \\ a_{21} & a_{22} & a_{23} \\ a_{31} & a_{32} & a_{33} \end{bmatrix} \tag{5.60}$$

and

$$|\mathbf{A}| = a_{11}\begin{vmatrix} a_{22} & a_{23} \\ a_{32} & a_{33} \end{vmatrix} - a_{12}\begin{vmatrix} a_{21} & a_{23} \\ a_{31} & a_{33} \end{vmatrix} + a_{13}\begin{vmatrix} a_{21} & a_{22} \\ a_{31} & a_{32} \end{vmatrix}. \tag{5.61}$$

The minor of an element a_{ij} of a determinant $|\mathbf{A}|$ is another determinant $|\mathbf{A}'_{ij}|$ obtained by deleting elements of the ith row and jth column of $|\mathbf{A}|$ (Figure 5.2). If $|\mathbf{A}|$ is 4×4, then all $|\mathbf{A}'_{ij}|$ are 3×3.

$$|\mathbf{A}| = \begin{vmatrix} a_{11} & a_{12} & a_{13} & a_{14} \\ a_{21} & a_{22} & a_{23} & a_{24} \\ a_{31} & a_{32} & a_{33} & a_{34} \\ a_{41} & a_{42} & a_{43} & a_{44} \end{vmatrix} \qquad \text{Minor } a_{32} = \begin{vmatrix} a_{11} & a_{13} & a_{14} \\ a_{21} & a_{23} & a_{24} \\ a_{41} & a_{43} & a_{44} \end{vmatrix}$$

Figure 5.2 Minor of a determinant

The cofactor of element a_{ij} of a determinant $|\mathbf{A}|$ is obtained by the product of the minor of the element with a plus or minus sign determined according to

$$c_{ij} = (-1)^{i+j} \left|\mathbf{A}'_{ij}\right|. \tag{5.62}$$

We now arrange the elements C_{ij} into a matrix of cofactors, **C**, where $\left|\mathbf{A}'_{ij}\right|$ denotes a minor of $|\mathbf{A}|$. The value of a determinant is simply equal to the sum of the products of each element of any row (or column) and its cofactor. By successively applying this process, the determinant is reduced to a computable expression consisting of the multiplication and summation of elements of the array.

If the determinant of a matrix is zero, then the matrix is singular and has no inverse.

There is an interesting geometric interpretation of the determinant of a 2x2 matrix when it has real number elements: It gives the oriented area of the parallelogram with vertices at (0, 0), (a_{11}, a_{12}), (a_{21}, a_{22}), and $(a_{11} + a_{21}, a_{12} + a_{22})$ (Figure 5.3). The oriented area is the same as the usual area, except that it is negative when the vertices are listed in clockwise order.

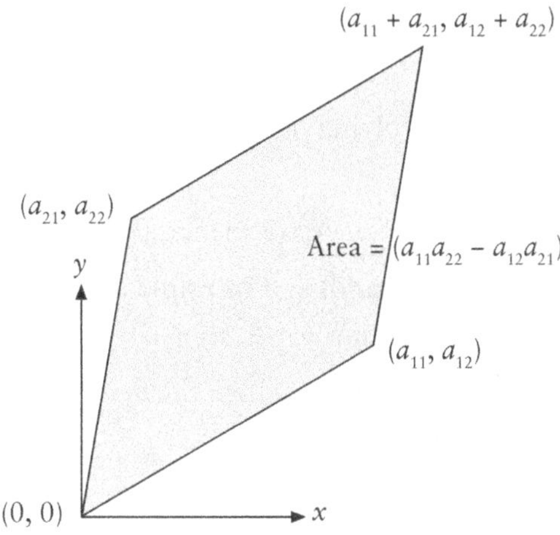

Figure 5.3 A determinant defining the area of a parallelogram

Here are some applications of determinants:

1. Determinants characterize invertible matrices (matrices with nonzero determinants, for example) and describe the solution to a system of linear equations.
2. They solve the eigenvalues of the matrix **A** through the characteristic polynomial $p(\lambda) = \det(\lambda\mathbf{I} - \mathbf{A})$, where **I** is the identity matrix of the same dimension as **A**. You may also see this expressed as

$$p(x) = \det(x\mathbf{I} - \mathbf{A}). \tag{5.63}$$

3. A determinant assigns a number to every sequence of n vectors in an n-dimensional space by using a square matrix whose columns are the given vectors. It happens that the sign of the determinant of a basis defines the notion of orientation in Euclidean spaces. The determinant of a set of vectors is positive if the vectors form a right-hand coordinate system and negative if a left-hand system.
4. The absolute value of the determinant of real vectors is equal to the volume of the parallelepiped spanned by those vectors.
5. The volume of any tetrahedron, given its vertices as vectors **a**, **b**, **c**, and **d**, is

$$\frac{1}{6} | \det(\mathbf{a}-\mathbf{b}, \mathbf{b}-\mathbf{c}, \mathbf{c}-\mathbf{d}) | . \tag{5.64}$$

5.10 Summary of Determinant Properties

As with matrix properties, create some examples of each determinant property to demonstrate its truth:

1. The determinant of a square matrix is equal to the determinant of its transpose: $|\mathbf{A}| = |\mathbf{A}^T|$.
2. Interchanging any two rows (or any two columns) of **A** changes the sign of $|\mathbf{A}|$.
3. If **B** is obtained by multiplying one row (or column) of **A** by a constant k, then $|\mathbf{B}| = k\,|\mathbf{A}|$.
4. If two rows (or columns) of **A** are identical, then $|\mathbf{A}| = 0$.
5. If **B** is derived from **A** by adding a multiple of one row (or column) of **A** to another row (or column) of **A**, then $|\mathbf{B}| = |\mathbf{A}|$.
6. If **A** and **B** are both $n \times n$ matrices, then the determinant of their product is $|\mathbf{AB}| = |\mathbf{A}||\mathbf{B}|$.
7. If every element of a row (or column) is zero, then the value of the determinant is zero.
8. If the determinant of a square matrix **A** is equal to 1, written mathematically as $|\mathbf{A}| = +1$, then it is orthogonal and proper. If $|\mathbf{A}| = -1$, it is orthogonal and improper. The orthogonal property is important in geometric transformations theory and application.
9. Adding a multiple of a row or column to another leaves the determinant unchanged.
10. Sylvester's determinant theorem states that for any $m \times n$ matrices A and B, $\det\left(\mathbf{I}_m + \mathbf{AB}^T\right) = \det\left(\mathbf{I}_n + \mathbf{B}^T\mathbf{A}\right)$.

5.11 Exercises

5.1. Given $\mathbf{A} = \begin{bmatrix} 7 & 4 & 4 \\ 9 & 1 & 3 \\ 0 & 2 & 5 \end{bmatrix}$ and $\mathbf{B} = \begin{bmatrix} 6 & 5 \\ 8 & 1 \\ 3 & 9 \end{bmatrix}$,

a. Find a_{23}.

b. Find a_{12}.

c. Find a_{31}.

d. Find b_{11}.

e. Find b_{32}.

f. What is the order of **A**?

g. What is the order of **B**?

h. Which matrix, if any, is a square matrix?

i. List the elements, in order, on the main diagonal of **A**.

j. Change a_{12}, a_{13}, and a_{23} so that **A** is a symmetric matrix.

5.2. Find the values of the following δ_{ij}:

a. $\delta_{3,2}$

b. $\delta_{1,4}$

c. $\delta_{3,3}$

d. $\delta_{7,10}$

e. $\delta_{1,1}$

5.3. Given $\mathbf{A} = \begin{bmatrix} 5 & 4 & 9 \\ 2 & 1 & 0 \\ 6 & 7 & 1 \end{bmatrix}$,

a. Change a_{12}, a_{13}, and a_{23} so that **A** becomes antisymmetric.

b. What other changes, if any, are necessary?

5.4. Write out the 2×2 null matrix.

5.5. Are the following matrices orthogonal, proper, or improper?

a. $\begin{bmatrix} \frac{\sqrt{2}}{2} & \frac{\sqrt{2}}{2} \\ -\frac{\sqrt{2}}{2} & \frac{\sqrt{2}}{2} \end{bmatrix}$

b. $\begin{bmatrix} 1 & \frac{1}{2} \\ 2 & 0 \end{bmatrix}$

c. $\begin{bmatrix} 3 & 1 \\ 5 & 2 \end{bmatrix}$

d. $\begin{bmatrix} -\frac{3}{2} & \frac{1}{2} \\ \frac{1}{2} & \frac{3}{2} \end{bmatrix}$

e. $\begin{bmatrix} 0 & 0 & 1 \\ 1 & 0 & 0 \\ 0 & -1 & 0 \end{bmatrix}$

f. $\begin{bmatrix} \frac{1}{2} & -\frac{\sqrt{3}}{2} & 0 \\ \frac{\sqrt{3}}{2} & \frac{1}{2} & 0 \\ 0 & 0 & 2 \end{bmatrix}$

g. $\begin{bmatrix} \frac{\sqrt{3}}{2} & \frac{\sqrt{3}}{4} & 1 \\ \frac{1}{2} & -\frac{3}{4} & -\frac{\sqrt{3}}{4} \\ 0 & \frac{1}{2} & -\frac{\sqrt{3}}{2} \end{bmatrix}$

5.6. Find $\begin{bmatrix} 0 & 0 & 1 \\ 1 & 0 & 0 \\ 0 & -1 & 0 \end{bmatrix}^{-1}$.

5.7. Show that the inverse of the orthogonal matrix **A** is an orthogonal matrix, where

$$\mathbf{A} = \begin{bmatrix} \frac{2}{3} & -\frac{2}{3} & \frac{1}{3} \\ \frac{1}{3} & \frac{2}{3} & \frac{2}{3} \\ \frac{2}{3} & \frac{1}{3} & -\frac{2}{3} \end{bmatrix}.$$

5.8. Compute the following determinants:

a. $\begin{vmatrix} 2 & 0 \\ -3 & 2 \end{vmatrix}$

b. $\begin{vmatrix} 1 & 2 \\ 4 & -5 \end{vmatrix}$

c. $\begin{vmatrix} 0 & 3 \\ 0 & 0 \end{vmatrix}$

d. $\begin{vmatrix} 1 & 2 \\ 2 & 4 \end{vmatrix}$

e. $\begin{vmatrix} 2 & 5 \\ -3 & 1 \end{vmatrix}$

5.9. Given $|\mathbf{A}| = \begin{vmatrix} 4 & 0 & -1 \\ 1 & 2 & 1 \\ -3 & 6 & 5 \end{vmatrix}$, compute the following minors and cofactors:

a. m_{11}
b. m_{21}
c. m_{31}
d. m_{22}
e. m_{12}
f. c_{11}
g. c_{21}
h. c_{31}
i. c_{22}
j. c_{12}

5.10. Compute $|\mathbf{A}|$ in the preceding exercise.

5.11. Compute the inverse of the following matrices, if one exists:

a. $\begin{bmatrix} 1 & 0 \\ 0 & 1 \end{bmatrix}$

b. $\begin{bmatrix} 3 & -1 & 2 \\ 1 & 2 & 1 \\ -2 & 1 & 3 \end{bmatrix}$

c. $\begin{bmatrix} 1 & 0 & 0 \\ 2 & 1 & 3 \\ 1 & 1 & 2 \end{bmatrix}$

d. $\begin{bmatrix} 3 & -1 & 2 \\ 1 & 2 & 1 \\ 3 & -1 & 2 \end{bmatrix}$

5.12. Find the eigenvalues and eigenvectors of the following matrices:

a. $\begin{bmatrix} 3 & 5 \\ 4 & 5 \end{bmatrix}$

b. $\begin{bmatrix} 1 & 2 \\ -2 & 5 \end{bmatrix}$

5.13. Find the characteristic equation, eigenvalues, and corresponding eigenvectors for the following matrices:

a. $\begin{bmatrix} 1 & 2 \\ 4 & 3 \end{bmatrix}$

b. $\begin{bmatrix} 2 & 0 & 0 \\ 0 & 1 & 0 \\ 0 & 0 & 3 \end{bmatrix}$

5.14. Prove that $|\mathbf{AB}| = |\mathbf{BA}| = |\mathbf{A}||\mathbf{B}| = |\mathbf{B}||\mathbf{A}|$ for any two square matrices **A** and **B**. (Note that this is true for square matrices of any order.)

6 Basis Vectors

Basis vectors and the coordinate systems they establish are the focus of this chapter. We can use them to construct the familiar orthonormal Cartesian system in Euclidean space, oblique systems, and more general affine systems, where the distance scale may be different along each of the three coordinate axes. Add a locating point, not necessarily the origin, to a set of basis vectors, and you have a frame.

6.1 Oblique Coordinate Systems

In a rectangular Cartesian coordinate system within Euclidean space, we can construct shapes with a compass and ruler, allowing us to produce right angles and circles and a distance scale that is the same in all directions. In nonorthogonal coordinate systems, we can measure the ratio of lengths along parallel lines, where the scale we use to measure distances may vary depending on the direction in which we take these measurements. We can produce ellipses in a nonorthogonal space, but not circles. Parallelism turns out to be an important invariant in nonorthogonal coordinate systems. The standard Cartesian coordinate system is almost too familiar. But there are advantages in certain situations to generalizing it. This is where basis vectors come in to play. They allow us to construct oblique and more general nonorthogonal coordinate systems, which make it easier to construct and analyze certain kinds of geometric models.

Here is one way to do this: First, remove the restriction that the principal planes and axes must be mutually perpendicular. Second, allow a different measuring scale for each axis. Next, assign an origin in this generalized coordinate system. We find the coordinates of any point in it as follows: Construct lines of projection from the point onto each axis so that each line is parallel to the plane formed by the other two axes. The intersection points on the axes yield the coordinates of the point, depending on the scale of each axis.

The simplest nonorthogonal coordinate system is the oblique coordinate system, where the unit of length is the same along all its axes. This last condition is not necessary in a general nonorthogonal coordinate system. Figure 6.1 compares orthogonal and oblique coordinate systems. The principal axes, E_1 and E_2, form an orthogonal Cartesian coordinate system.

F_1 and F_2 are the principal axes of a nonorthogonal coordinate system. The point P has components e_1 and e_2 in the E_1, E_2 system and coordinates f_1 and f_2 in the F_1, F_2 oblique coordinate system. Later we will see how to use vectors to construct nonorthogonal systems. That will open our way to study basis vector coordinate systems and frames and the notion of barycentric coordinates.

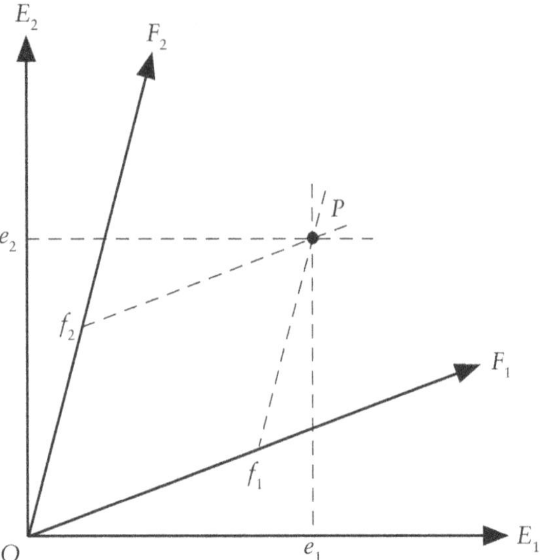

Figure 6.1 Coordinates of a point in orthogonal and oblique coordinate systems, sharing a common origin

6.2 Basis Vector Basics

A set of three linearly independent vectors $\mathbf{e}_1$, $\mathbf{e}_2$, and $\mathbf{e}_3$ forms a coordinate system. The vectors are analogous to **i**, **j**, and **k** in the orthonormal Cartesian system. We call them basis vectors, and we can express any position vector **r** as a linear combination of them. Let's see how this works (Figure 6.2). From a common point O, the origin, three families of parallel lines form a coordinate system. The lines X_1, X_2, and X_3 concurrent at O and collinear with $\mathbf{e}_1$, $\mathbf{e}_2$, and $\mathbf{e}_3$, respectively, define the coordinate axes.

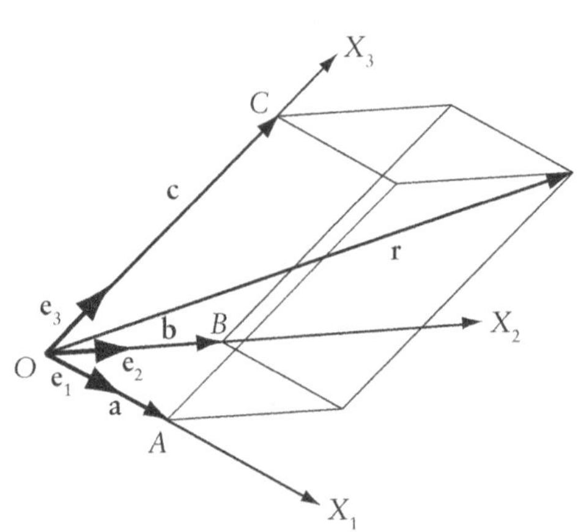

Figure 6.2 A three-dimensional basis

For any point (position vector) **r** in this system, we find coordinates (components) as follows: Construct a parallelepiped with O at one vertex, **r** as a body diagonal, and concurrent edges lying collinear with the basis vectors. The three directed line segments corresponding to edges OA, OB, and OC define the vector components of **r** in this system. Denote these as **a**, **b**, and **c**, respectively. The very nature of this construction technique ensures that the parallelogram law of vector addition applies. Thus,

$$\mathbf{r} = \mathbf{a} + \mathbf{b} + \mathbf{c}. \tag{6.1}$$

What are the coordinates of **r** relative to this basis? To answer this we begin with these definitions: Let

$$r_1 = \frac{|\mathbf{a}|}{|\mathbf{e}_1|}, \quad r_2 = \frac{|\mathbf{b}|}{|\mathbf{e}_2|}, \quad r_3 = \frac{|\mathbf{c}|}{|\mathbf{e}_3|}. \tag{6.2}$$

Remember, we do not insist that basis vectors must be unit vectors. Now, using these definitions, we rewrite Equation 6.1 to obtain

$$\mathbf{r} = r_1\mathbf{e}_1 + r_2\mathbf{e}_2 + r_3\mathbf{e}_3, \tag{6.3}$$

where we identify r_1, r_2, and r_3 as the components or coordinates of **r** with respect to the frame of reference defined by the basis vectors $\mathbf{e}_1$, $\mathbf{e}_2$, and $\mathbf{e}_3$. Sometimes we speak of these as the parallel coordinates of the point **r**. These coordinates coincide with the coordinates of the affine three-dimensional space if the basis vectors are unit vectors. Notice that this system of basis vectors, as shown in Figure 6.2, is a right-hand one, but it could just as well be left hand.

We have just constructed a nonorthogonal coordinate system of the most general sort. As mentioned earlier, such a system may offer computational or other advantages if the principal vectors defining a particular problem are not orthogonal. Now let's look at how we do geometry in an oblique system.

First, let vectors $\mathbf{e}_1$ and $\mathbf{e}_2$ define an oblique coordinate system in the xy plane; then every point in the plane has a position vector given by

$$\mathbf{p} = \alpha\mathbf{e}_1 + \beta\mathbf{e}_2, \tag{6.4}$$

where the oblique coordinates of a point are (α, β) (Figure 6.3). The magnitudes of $\mathbf{e}_1$ and $\mathbf{e}_2$ establish the scaling along the two axes.

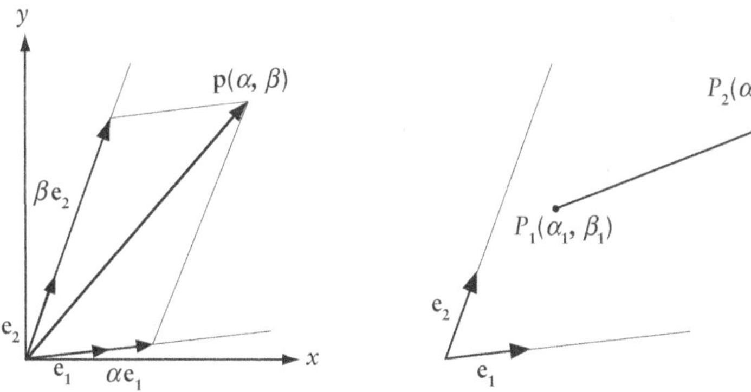

Figure 6.3 Oblique coordinate system

Using the properties of similar triangles, we write the equation of a line through points $P_1(\alpha_1, \beta_1)$ and $P_2(\alpha_2, \beta_2)$ as

$$(\alpha - \alpha_1)(\beta_2 - \beta_1) = (\beta - \beta_1)(\alpha_2 - \alpha_1). \quad (6.5)$$

For example, the equation of the line through $P_1(1,0)$ and $P_2(0,1)$ is $\alpha + \beta = 1$.

The x, y coordinates of any point with oblique coordinates α, β are, in matrix terms,

$$\begin{bmatrix} x \\ y \end{bmatrix} = \begin{bmatrix} e_{1,x} & e_{2,x} \\ e_{1,y} & e_{2,y} \end{bmatrix} \begin{bmatrix} \alpha \\ \beta \end{bmatrix}, \quad (6.6)$$

where $(e_{1,x}, e_{1,y})$ and $(e_{2,x}, e_{2,y})$ are the orthogonal components of $\mathbf{e}_1$ and $\mathbf{e}_2$.

After performing the indicated matrix multiplication, we obtain

$$\begin{aligned} x &= \alpha e_{1,x} + \beta e_{2,x}, \\ y &= \alpha e_{1,y} + \beta e_{2,y}. \end{aligned} \quad (6.7)$$

Finally, we find that the square of the distance between two points $P_1(\alpha_1, \beta_1)$ and $P_2(\alpha_2, \beta_2)$ is

$$\begin{aligned} \left(\overline{P_1P_2}\right)^2 &= |\mathbf{p}_2 - \mathbf{p}_1|^2 \\ &= |(\alpha_2 - \alpha_1)\mathbf{e}_1 + (\beta_2 - \beta_1)\mathbf{e}_2|^2 \\ &= (\alpha_2 - \alpha_1)^2 \mathbf{e}_1 \bullet \mathbf{e}_1 + 2(\alpha_2 - \alpha_1)(\beta_2 - \beta_1)\mathbf{e}_1 \bullet \mathbf{e}_2 + (\beta_2 - \beta_1)^2 \mathbf{e}_2 \bullet \mathbf{e}_2. \end{aligned} \quad (6.8)$$

If $\mathbf{e}_1$ and $\mathbf{e}_2$ are mutually perpendicular unit vectors, then

$$\mathbf{e}_1 \bullet \mathbf{e}_1 = \mathbf{e}_2 \bullet \mathbf{e}_2 = 1 \quad (6.9)$$

and

$$\mathbf{e}_1 \bullet \mathbf{e}_2 = 0, \quad (6.10)$$

and the expression for distance simplifies to the familiar Pythagorean theorem of rectangular Cartesian coordinates.

6.3 Change of Basis

There is no limit to the number of basis systems we can construct in a space. Every basis of a space determines a unique coordinate system and a corresponding set of coordinates for any point in the space. So it is natural to ask: How do we change from one basis to another? We might also ask: Given the components of a vector relative to one basis, what are its components relative to another? We simply apply linear transformations. This reveals another useful property of bases: Different bases can represent the same vector, just so they are the same dimension.

Given a basis $\mathbf{e}_1, \mathbf{e}_2$, we change to another basis $\bar{\mathbf{e}}_1, \bar{\mathbf{e}}_2$ using a set of linear transformation equations (Figure 6.4):

$$\begin{aligned}\bar{\mathbf{e}}_1 &= a\mathbf{e}_1 + b\mathbf{e}_2,\\ \bar{\mathbf{e}}_2 &= c\mathbf{e}_1 + d\mathbf{e}_2,\end{aligned} \tag{6.11}$$

where $ad - bc \neq 0$.

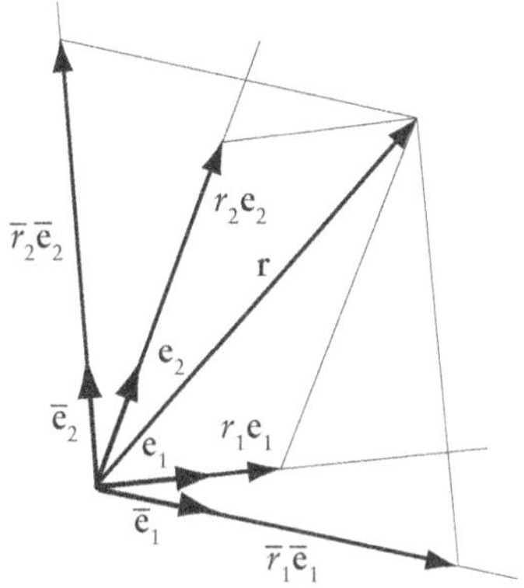

Figure 6.4 Change of basis in two dimensions

The equation expressing the components of any vector $\mathbf{r}$ in basis $\mathbf{e}_1, \mathbf{e}_2$ is

$$\mathbf{r} = r_1\mathbf{e}_1 + r_2\mathbf{e}_2. \tag{6.12}$$

This is what we might expect by analogy with Equation 6.3.

If we solve Equation 6.11 for $\mathbf{e}_1$ and $\mathbf{e}_2$, we obtain

$$\begin{aligned}\mathbf{e}_1 &= \frac{d}{\Delta}\bar{\mathbf{e}}_1 - \frac{b}{\Delta}\bar{\mathbf{e}}_2,\\ \mathbf{e}_2 &= -\frac{c}{\Delta}\bar{\mathbf{e}}_1 + \frac{a}{\Delta}\bar{\mathbf{e}}_2,\end{aligned} \tag{6.13}$$

where $\Delta = ad - bc$.

Substitute from Equation 6.9 into 6.8 for $\mathbf{e}_1$ and $\mathbf{e}_2$ and rearrange terms as necessary, so that

$$\mathbf{r} = \left(\frac{d}{\Delta}r_1 - \frac{c}{\Delta}r_2\right)\bar{\mathbf{e}}_1 + \left(-\frac{b}{\Delta}r_1 + \frac{a}{\Delta}r_2\right)\bar{\mathbf{e}}_2. \tag{6.14}$$

Because of the objective (invariant) nature we attribute to $\mathbf{r}$ or any vector, we expect to describe the components of $\mathbf{r}$ in terms of the new basis $\bar{\mathbf{e}}_1, \bar{\mathbf{e}}_2$ and so we write

$$\mathbf{r} = \bar{r}_1\bar{\mathbf{e}}_1 + \bar{r}_2\bar{\mathbf{e}}_2. \tag{6.15}$$

Now compare Equations 6.14 and 6.15. It must be true that

$$\begin{aligned}\bar{r}_1 &= \frac{d}{\Delta}r_1 - \frac{c}{\Delta}r_2,\\ \bar{r}_2 &= -\frac{b}{\Delta}r_1 + \frac{a}{\Delta}r_2.\end{aligned} \tag{6.16}$$

Take the matrix **B** of coefficients that transforms the $\bar{\mathbf{e}}$ basis to the **e** basis and compare it with the matrix **C** of coefficients that transforms the components of **r** from the **e** basis to the $\bar{\mathbf{e}}$ basis:

$$\mathbf{B} = \begin{bmatrix} \frac{d}{\Delta} & -\frac{b}{\Delta} \\ -\frac{c}{\Delta} & \frac{a}{\Delta} \end{bmatrix} \qquad \mathbf{C} = \begin{bmatrix} \frac{d}{\Delta} & -\frac{c}{\Delta} \\ -\frac{b}{\Delta} & \frac{a}{\Delta} \end{bmatrix}. \tag{6.17}$$

We immediately notice that **C** is the transpose of **B**, or $\mathbf{C} = \mathbf{B}^T$, so that their elements correspond accordingly to

$$c_{ij} = b_{ji}. \tag{6.18}$$

If we denote the matrix that transforms the **e** basis into the $\bar{\mathbf{e}}$ basis as **A**, then clearly $\mathbf{B} = \mathbf{A}^{-1}$. This gives us

$$\mathbf{C} = \left[\mathbf{A}^{-1}\right]^T. \tag{6.19}$$

The relationships expressed in Equations 6.11, 6.13, and 6.16 are deceptively simple but, nonetheless, very powerful. They show how the equations transforming basis vectors relate to those transforming the vector components.

6.4 Reciprocal Basis Vectors

Two sets of basis vectors, $(\mathbf{e}_1, \mathbf{e}_2, \mathbf{e}_3,)$ and $(\mathbf{e}^1, \mathbf{e}^2, \mathbf{e}^3)$, are reciprocal if and only if the following relations are true:

$$\begin{matrix} \mathbf{e}_1 \bullet \mathbf{e}^1 = 1 & \mathbf{e}_1 \bullet \mathbf{e}^2 = 0 & \mathbf{e}_1 \bullet \mathbf{e}^3 = 0 \\ \mathbf{e}_2 \bullet \mathbf{e}^1 = 0 & \mathbf{e}_2 \bullet \mathbf{e}^2 = 1 & \mathbf{e}_2 \bullet \mathbf{e}^3 = 0 \\ \mathbf{e}_3 \bullet \mathbf{e}^1 = 0 & \mathbf{e}_3 \bullet \mathbf{e}^2 = 0 & \mathbf{e}_3 \bullet \mathbf{e}^3 = 1. \end{matrix} \tag{6.20}$$

The superscript index has significance beyond its most apparent role here, namely that of helping to denote a distinction between the two sets of basis vectors. We will discuss this point later.

The Kronecker delta, δ_i^j, allows us to compress these nine equations into the much more compact form of

$$\mathbf{e}_i \bullet \mathbf{e}^j = \delta_i^j. \tag{6.21}$$

Note that $\delta_i^j = 1$ if $i = j$; otherwise $\delta_i^j = 0$. For example, the equations $\mathbf{e}_1 \bullet \mathbf{e}^2 = 0$ and $\mathbf{e}_1 \bullet \mathbf{e}^3 = 0$ tell us that $\mathbf{e}_1$ is perpendicular to both $\mathbf{e}^2$ and $\mathbf{e}^3$.

Furthermore, we notice that a set of basis vectors and the set of their reciprocals are identical if

$$\mathbf{e}_i \bullet \mathbf{e}_j = \partial_{ij}. \tag{6.22}$$

There is no reason to use superscripts or to distinguish between reciprocal sets of basis vectors that satisfy this condition, since they are orthonormal unit vectors. The orthonormal Cartesian basis system has this property.

Let us look at what happens in a two-dimensional vector space (Figure 6.5). We begin with two unit vectors $\hat{e}_1, \hat{e}_2$ forming a basis on the x^1, x^2 axes. We then construct a new set of axes x_1, x_2 such that x_1 is perpendicular to x^2 and x_2 is perpendicular to x^1. This means that

$$\begin{aligned} &\hat{e}_1 \bullet \hat{e}^2 = 0, \\ &\hat{e}_2 \bullet \hat{e}^1 = 0, \\ &\hat{e}_1 \bullet \hat{e}^1 = |\hat{e}_1||\hat{e}^1| \sin\theta. \end{aligned} \tag{6.23}$$

From the figure we see that $\sin\theta = |\hat{e}_1|/|\hat{e}^1|$, so that

$$\begin{aligned} \hat{e}_1 \bullet \hat{e}^1 &= |\hat{e}_1||\hat{e}^1| \frac{|\hat{e}_1|}{|\hat{e}^1|} \\ &= |\hat{e}_1|^2 \\ &= 1, \end{aligned} \tag{6.24}$$

and similarly $\hat{e}_2 \bullet \hat{e}^2 = 1$. Thus $\hat{e}_1, \hat{e}_2$ and $\hat{e}^1, \hat{e}^2$ are reciprocal sets of basis vectors.

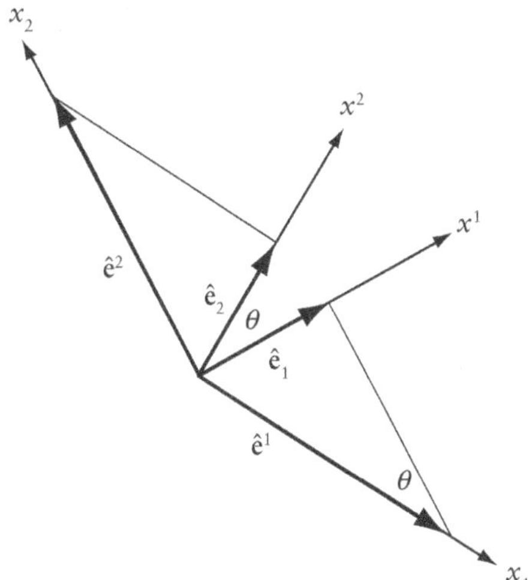

Figure 6.5 Reciprocal basis vectors

Finally, notice that the scale of the $|\hat{e}_i|$ is different from that of the $|\hat{e}^j|$. Since we began with the assumption that $\hat{e}_1$ and $\hat{e}_2$ are unit vectors, it should be clear that $|\hat{e}^j| > 1$; to put it another way, we see that $|\hat{e}^1| = 1/\sin\theta > 1$.

6.5 Orthogonal Basis Vectors and Matrices

If the angle between two vectors is a right angle, then the vectors are mutually orthogonal. Recall the condition that characterizes orthogonal vectors in three-dimensional space:

$$\mathbf{p} \bullet \mathbf{q} = 0, \tag{6.25}$$

or

$$p_x q_x + p_y q_y + p_z q_z = 0. \tag{6.26}$$

The most common system of basis vectors is one in which each basis vector in it is mutually perpendicular to the others, so that $\mathbf{e}_i \bullet \mathbf{e}_j = \delta_{ij}$, an orthogonal system corresponding to an orthonormal Cartesian system (Figure 6.6). If it is also true that $|\mathbf{e}_i| = 1$, then it is a system of orthonormal basis vectors.

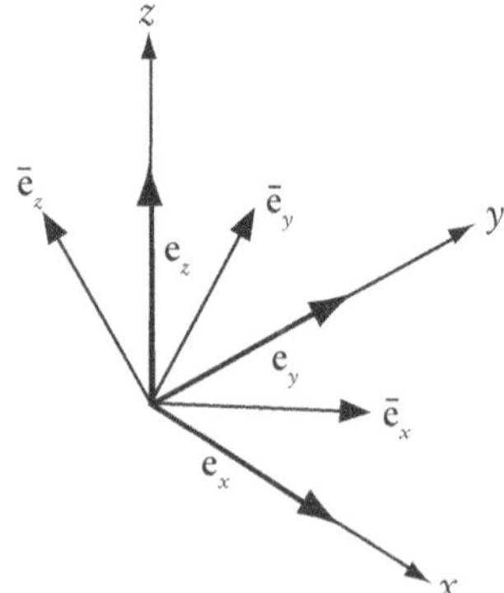

Figure 6.6 Orthonormal basis vectors

If $\mathbf{e}_i$ is an orthonormal basis, and if a_{ij} represents elements of the transformation matrix from this basis to another, $\bar{\mathbf{e}}_i$, then a necessary and sufficient condition for the $\bar{\mathbf{e}}_i$ basis to be orthonormal is that a_{ij} must be an orthogonal matrix. Recall that orthogonal vectors are mutually perpendicular without restriction on their magnitudes. Orthonormal vectors are orthogonal with the restriction that they are also unit vectors.

Given a real, square matrix $\mathbf{A}$ of order n, if $\mathbf{AA}^{\mathrm{T}} = \mathbf{I}$, then $\mathbf{A}^{\mathrm{T}} = \mathbf{A}^{-1}$ and $\mathbf{A}$ is an orthonormal matrix (review Section 5.5). Let a_{ij} denote the elements of $\mathbf{A}$; then if $\mathbf{A}$ is orthonormal, we have

$$\sum_{k=1}^{n} a_{ik} a_{jk} = \delta_{ij}. \tag{6.27}$$

6.6 Frames

We define a frame as an ordered basis plus a (reference) point, which we may (with care) treat as a vector. There are other definitions, usually minor variations of this. Let's try some descriptions and applications and not worry too much more about a rigorous mathematical definition.

First, a frame is a coordinate system. It can be orthogonal or not. In practical terms, we assume it is embedded in a space of the same dimension and often with other frames floating around. In three-dimensional space, three linearly independent vectors define the frame coordinate system, along with an associated point locating the frame within the embedding space, with respect to some other point or origin. (*Note:* If none of the vectors is parallel to the others, then they are linearly independent.) We will mostly encounter frames in ordinary three-dimensional Euclidean space containing a Cartesian coordinate system, and that is how our discussion will proceed. However, frames are also right at home in affine space, which generalizes certain properties of Euclidean space . . . there is no distinct or natural origin, no vector has a fixed origin, and more.

The way we specify frames determines how we transform the definition and location of geometric objects relative to one frame into a new definition relative to another frame. For example, we define a 3D solid in its own local or object frame and later place it in a world or scene coordinate system. Another example is a virtual camera or viewpoint, which has its local coordinate system controlling its orientation, placed within a scene coordinate system. So where you find a frame, you will also find transformations lurking. They are all over the place when it comes to 3D model building, animation, and rendering.

Here is how we'll describe, apply, and illustrate a frame (Figure 6.7). We start with the embedding system (two dimensions for simplicity, but easily expanded to three), a point $\mathbf{p}_0$ in it, and basis vectors $\mathbf{e}_1$ and $\mathbf{e}_2$. We can specify any point $\mathbf{p}$ as

$$\mathbf{p} = \mathbf{p}_0 + c_1\mathbf{e}_1 + c_2\mathbf{e}_2, \tag{6.28}$$

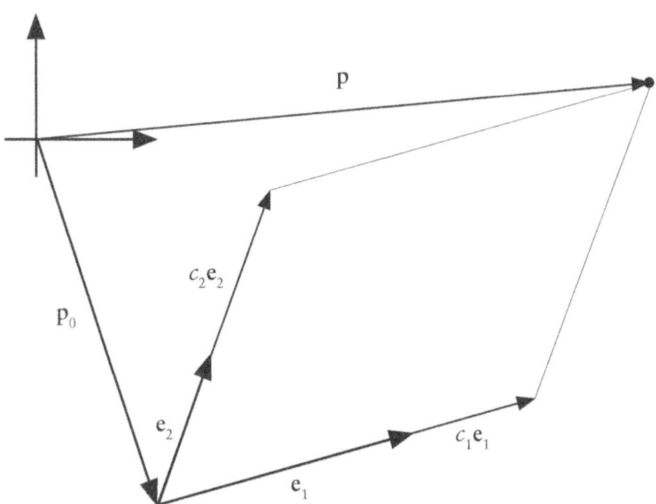

Figure 6.7 A two-dimensional frame

where $\mathbf{p}_0$ is the origin of the frame, and c_1 and c_2 are the coordinates of $\mathbf{p}$ relative to the $\mathbf{e}_1$ and $\mathbf{e}_2$ basis of the frame. Complete the parallelogram as suggested in the figure to see the geometric interpretation of the vector addition expressed above. Notice that this is not an orthogonal Cartesian frame.

Recall that the orthonormal Cartesian frame in two dimensions has basis vectors

$$\begin{aligned} \mathbf{e}_1 &= \begin{bmatrix} 1 & 0 \end{bmatrix} \\ \mathbf{e}_2 &= \begin{bmatrix} 0 & 1 \end{bmatrix}, \end{aligned} \tag{6.29}$$

with $\mathbf{p}_0 = [\,0 \quad 0\,]$.

A similar frame in three dimensions has basis vectors

$$\begin{aligned} \mathbf{e}_1 &= [\,1 \quad 0 \quad 0\,], \\ \mathbf{e}_2 &= [\,0 \quad 1 \quad 0\,], \\ \mathbf{e}_3 &= [\,0 \quad 0 \quad 1\,], \end{aligned} \tag{6.30}$$

with $\mathbf{p}_0 = [\,0 \quad 0 \quad 0\,]$.

7 Barycentric Coordinate Systems

Barycentric coordinate systems are unusual: They do not have an origin, principal axes, or principal planes . . . at least not in the customary sense. But a barycentric coordinate system is good at locating points within a triangle, which has application to shading polyhedra in rendering applications. This chapter presents the mathematics of these systems. And, of course, vectors play a significant role.

7.1 Origins

The term "barycenter" is borrowed from physics, where it is synonymous with "center of mass" or "center of gravity," and is analogous to the term "centroid" or "geometric center" in geometry (Figure 7.1). The mathematics and geometry of barycentric coordinate systems derive from these concepts. Geometric modeling, game design, collision detection, and many other computational geometry problems make use of barycentric coordinate systems to simplify solutions and speed up computation time. Barycentric coordinates are particularly useful for certain computations involving polyhedra and their polygonal faces, particularly when triangular faces approximate the surface of a geometric model.

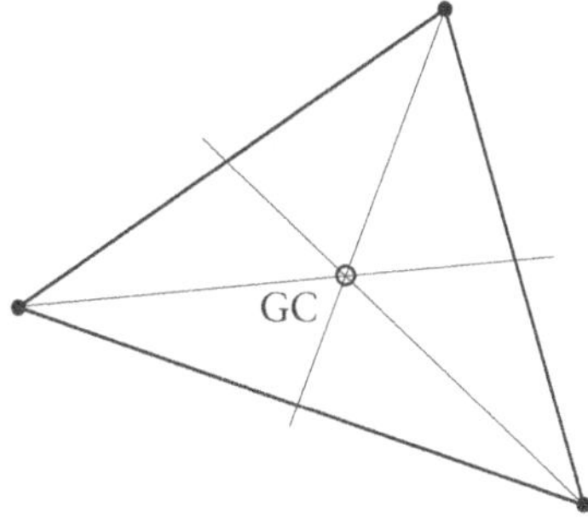

Figure 7.1 Geometric center of a triangle

7.2 Barycentric Coordinates

Let's start with a set of points in two-, three-, or n-dimensional space: $\mathbf{p}_1, \mathbf{p}_2, \ldots, \mathbf{p}_n$. We'll impose a restriction on them, whose full development is outside the scope of our work here; the sets of points we will work with define the vertices of a simplex.

Without diving into advanced math theory, here is a simple way (no pun intended) to think of a simplex: In two-dimensional space or three-dimensional space, it takes three noncoincident points to define a finite planar region. Connecting the points with three lines that define a triangle does this. Four noncoincident points in three-dimensional space suffice to define the vertices of a polyhedron, enclosing a finite volume . . . a tetrahedron.

We can continue this process into higher dimensions, creating simplexes within them that enclose finite regions. Let's see how this works in practice. A single point, the two endpoints of a line segment, the three vertex points of a triangle, and the four vertex points of a tetrahedron each define a simplex, the simplex being the point, the line segment, the triangle, and the tetrahedron, respectively.

We begin by defining a point in a system of barycentric coordinates as a linear combination of a set of given points. For example, for n points we have

$$\mathbf{p} = b_1\mathbf{p}_1 + b_2\mathbf{p}_2 + \ldots + b_n\mathbf{p}_n. \tag{7.1}$$

The coefficients $b_1, b_2, \ldots, b_n$ are the barycentric coordinates of point $\mathbf{p}$ with respect to $\mathbf{p}_1, \mathbf{p}_2, \ldots, \mathbf{p}_n$. We put these two restrictions on them:

$$b_1 + b_2 + \ldots + b_n = 1 \tag{7.2}$$

and

$$0 \le b_1, b_2, \ldots, b_n \le 1. \tag{7.3}$$

Another way of writing Equation 7.3 is

$$\sum_1^n b_i = 1. \tag{7.4}$$

Notice that Equation 7.3 restricts each barycentric coordinate to the interval from 0 to 1, and the angle bracket symbol $\langle\ \rangle$ distinguishes barycentric coordinates from ordinary Cartesian coordinates. The fields of statistics and linear algebra (as well as several others) also use this symbol. Here it will indicate and enclose an ordered pair, triple, or n-tuple, for example $\langle b_1, b_2, b_3 \rangle$.

The barycentric coordinate system, then, is a coordinate system that locates points within a simplex. (Actually, points can be located outside these simplexes, too. But unless it is a test for containment or not, it is the points inside that are important.)

7.3 Relative to Two Points

For a simplex defined by two points, we have

$$\mathbf{p} = b_1\mathbf{p}_1 + b_2\mathbf{p}_2 \tag{7.5}$$

and if $\mathbf{b}_1 = 0.33$ and $\mathbf{b}_2 = 0.67$, then

$$b_1 + b_2 = 1. \tag{7.6}$$

The barycentric coordinates of $\mathbf{p}$ are $\langle b_1, b_2 \rangle$, with respect to $\mathbf{p}_1$ and $\mathbf{p}_2$. Think of the coordinates as weighting or influence factors. If $b_1 > b_2$, then $\mathbf{p}_1$ contributes more to the value of $\mathbf{p}$ than does $\mathbf{p}_2$.

Here is an example (Figure 7.2). This one is in two dimensions only because it is easier to illustrate. But it could just as well be in three dimensions.

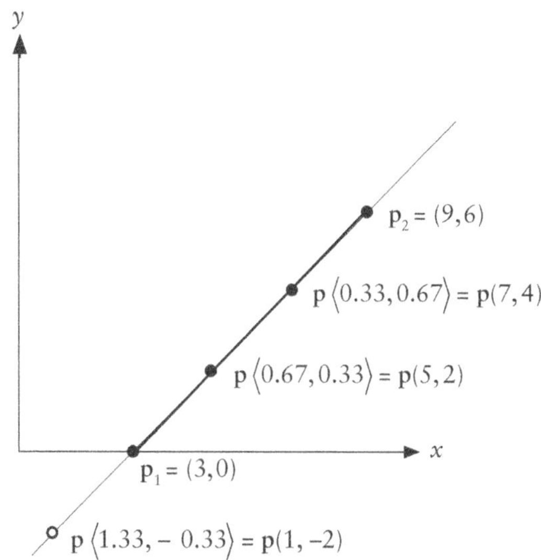

Figure 7.2 Barycentric coordinates of points between two given points

Let $\mathbf{p}_1 = (3, 0)$ and $\mathbf{p}_2 = (9, 6)$. If $b_1 = 0.67$ and $b_2 = 0.33$, then

$$\begin{aligned} \mathbf{p}\langle 0.67, 0.33 \rangle &= 0.67(3, 0) + 0.33(9, 6) \\ &= (5, 2) \end{aligned} \tag{7.7}$$

and if $\mathbf{b}_1 = 0.33$ and $\mathbf{b}_2 = 0.67$, then

$$\begin{aligned} \mathbf{p}\langle 0.33, 0.67 \rangle &= 0.33(3, 0) + 0.67(9, 6) \\ &= (7, 4). \end{aligned} \tag{7.8}$$

Notice that $\mathbf{p}\langle 0.67, 0.33 \rangle$ lies on the line between $\mathbf{p}_1$ and $\mathbf{p}_2$ and closer to $\mathbf{p}_1$, while $\mathbf{p}\langle 0.33, 0.67 \rangle$ also lies on that line but closer to $\mathbf{p}_2$. In fact any legitimate combination of barycentric coordinates $\langle b_1, b_2 \rangle$ applied to the set $\mathbf{p}_1$ and $\mathbf{p}_2$ produces a point on the line between them. So here is an example of a point that violates the restriction $0 \le b_1, b_2 \le 1$:

$$\mathbf{p}\langle 1.33, -0.33 \rangle = \mathbf{p}(1, -2). \tag{7.9}$$

The point is on the line but not on the segment between $\mathbf{p}_1$ and $\mathbf{p}_2$. The barycentric coordinates $\langle 1, 0 \rangle$ and $\langle 0, 1 \rangle$ produce the endpoints themselves, $\mathbf{p}_1$ and $\mathbf{p}_2$, respectively.

We take advantage of the relationship $b_1 + b_2 = 1$ or $b_1 = 1 - b_2$ to produce

$$\mathbf{p}\langle b_1, b_2\rangle = \mathbf{p}_1 + b_2(\mathbf{p}_2 - \mathbf{p}_1), \tag{7.10}$$

where $0 \le b_2 \le 1$. If we treat b_2 as an independent variable, then this is the equation of a line between $\mathbf{p}_1$ and $\mathbf{p}_2$.

Can the above process be reversed? Given any point on the line segment, can we find its barycentric coordinates? Yes, and for two points, $\mathbf{p}_1$ and $\mathbf{p}_2$, it is quite easy. Because $\mathbf{p}\langle b_1, b_2\rangle = \mathbf{p}_1 + b_2(\mathbf{p}_2 - \mathbf{p}_1)$, we solve for b_2 and then obtain b_1 from $b_1 = 1 - b_2$. Doing this we find

$$b_1 = 1 - \frac{x - x_1}{x_2 - x_1} \tag{7.11}$$

and

$$b_2 = \frac{x - x_1}{x_2 - x_1} \tag{7.12}$$

We can obtain the same results using ratios of the y coordinates, because a barycentric coefficient applies over all the coordinates of a point.

7.4 Relative to a Triangle

The concepts we've developed so far for two points also apply to three points, and it's not much more complicated. In Euclidean space, three points are always coplanar, and define a triangle, a two-dimensional simplex. Barycentric coordinates on a triangle become important when we deal with polygonized surfaces of geometric models. We will work with three points in the xy plane, again to simplify the illustrations and shorten the mathematical notations. Using the barycentric coordinates b_1, b_2, and b_3, we represent any point $\mathbf{p}$ in the triangle as

$$\mathbf{p} = b_1\mathbf{p}_1 + b_2\mathbf{p}_2 + b_3\mathbf{p}_3, \tag{7.13}$$

where

$$b_1 + b_2 + b_3 = 1 \tag{7.14}$$

and

$$0 \le b_1, b_2, b_3 \le 1. \tag{7.15}$$

Certain combinations of values of the barycentric coordinates b_1, b_2, and b_3 produce points on the vertices and edges of the triangle (Figure 7.3):

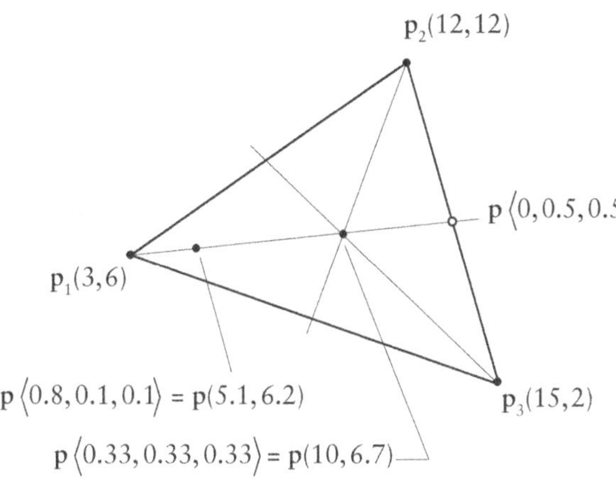

Figure 7.3 Barycentric coordinates of points in a triangle

If $b_1 = 1$ and $b_2 = b_3 = 0$, then $\mathbf{p} = \mathbf{p}_1$.
If $b_1 = 0$ and $b_2 = 1$ and $b_3 = 0$, then $\mathbf{p} = \mathbf{p}_2$.
If $b_1 = 0$ and $b_2 = 0$ and $b_3 = 1$, then $\mathbf{p} = \mathbf{p}_3$.
If $b_1 = 0$ and $0 \le b_2, b_3 \le 1$, then $\mathbf{p}$ is on $\mathbf{p}_2\mathbf{p}_3$.
If $b_2 = 0$ and $0 \le b_1, b_3 \le 1$, then $\mathbf{p}$ is on $\mathbf{p}_1\mathbf{p}_3$.
If $b_3 = 0$ and $0 \le b_1, b_2 \le 1$, then $\mathbf{p}$ is on $\mathbf{p}_1\mathbf{p}_2$.

Notice, again, that all the barycentric coordinates listed above have either positive or zero value.

There is more to see in Figure 7.3. The line through $\mathbf{p}_1$, $\mathbf{p}\langle 0.8, 0.1, 0.1\rangle$, and the centroid $\mathbf{p}\langle 0.33,\ 0.33,\ 0.33\rangle$ intersects the line through $\mathbf{p}_2$ and $\mathbf{p}_3$ at its midpoint, $\mathbf{p}\langle 0, 0.5, 0.5\rangle$. What does this suggest? We see that $b_2 = b_3$ at all points along a line through the centroid and the midpoint of a line segment connecting $\mathbf{p}_2$ and $\mathbf{p}_3$. We also see that when the barycentric coordinates are equal, $\langle 0.33, 0.33, 0.33\rangle$, they produce the point at the geometric center of the triangle (Figure 7.4).

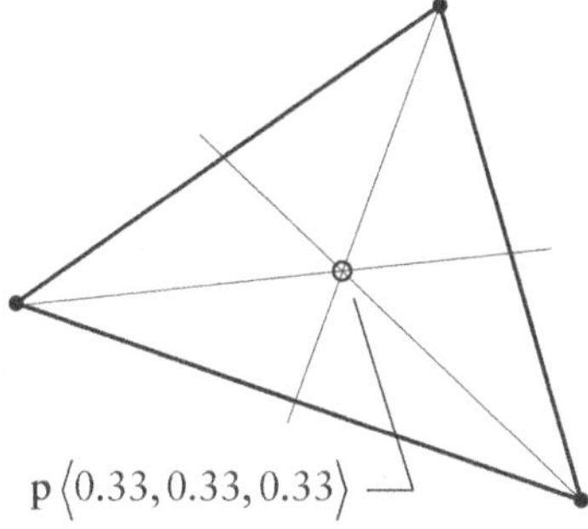

Figure 7.4 Barycentric coordinates of the geometric center of a triangle

Now look at the barycentric coordinates for the special case of an equilateral triangle (Figure 7.5). Find the barycentric coordinates of the vertices of the triangle in Figure 7.3, and compare to Figure 7.5.

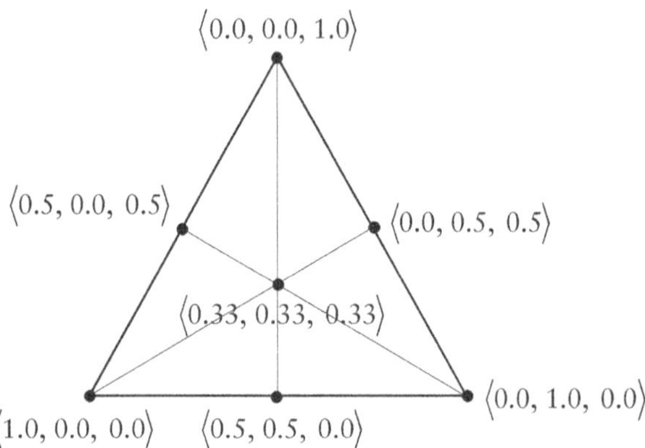

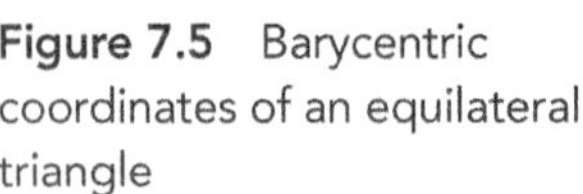
Figure 7.5 Barycentric coordinates of an equilateral triangle

Let's ask the same question we posed for the line segment, above: Can the process be reversed? Can we find the barycentric coordinates of a point in the triangle simplex? Again, the answer is yes. Here's how.

We know that the coordinates are

$$\begin{aligned} x &= b_1x_1 + b_2x_2 + b_3x_3, \\ y &= b_1y_1 + b_2y_2 + b_3y_3, \end{aligned} \tag{7.16}$$

with the conditions

$$b_1 + b_2 + b_3 = 1, \tag{7.17}$$

or alternatively,

$$b_3 = 1 - b_1 - b_2. \tag{7.18}$$

Using some simple algebra, we eliminate b_3 and rearrange to find

$$\begin{aligned} b_1(x_1 - x_3) + b_2(x_2 - x_3) &= (x - x_3), \\ b_1(y_1 - y_3) + b_2(y_2 - x_3) &= (y - y_3). \end{aligned} \tag{7.19}$$

There are several ways to solve these two equations in two unknowns: b_1 and b_2. There is, of course, the brute force and rather tedious way using elementary algebra that becomes completely intractable for a higher-order simplex, that is, for b_n when $n \geq 3$. The solution is much simpler and more general if we apply matrix algebra. (You can return to Chapters 4 and 5 on matrices for a review.)

We construct the following matrices using the elements of Equation 7.19:

$$\mathbf{A} = \begin{bmatrix} (x_1 - x_3) & (x_2 - x_3) \\ (y_1 - y_3) & (y_2 - y_3) \end{bmatrix}, \ \mathbf{B} = \begin{bmatrix} b_1 \\ b_2 \end{bmatrix}, \text{ and } \mathbf{P} = \begin{bmatrix} (x - x_3) \\ (y - y_3) \end{bmatrix}. \tag{7.20}$$

Now we write the matrix equation that is the equivalent of Equation 7.20:

$$\mathbf{AB} = \mathbf{P}. \tag{7.21}$$

Solve this for **P**:

$$\mathbf{B} = \mathbf{A}^{-1}\mathbf{P}, \tag{7.22}$$

or

$$\begin{bmatrix} b_1 \\ b_2 \end{bmatrix} = \begin{bmatrix} (x_1 - x_3) & (x_2 - x_3) \\ (y_1 - y_3) & (y_2 - y_3) \end{bmatrix}^{-1} \begin{bmatrix} (x - x_3) \\ (y - y_3) \end{bmatrix}. \tag{7.23}$$

When we do the indicated matrix math, we find our solution:

$$\begin{aligned} b_1 &= \frac{(x_2 - x_3)(x - x_3) + (x_3 - x_2)(y - y_3)}{(y_2 - y_3)(x_1 - x_3) + (x_3 - x_2)(y_1 - y_3)}, \\ b_2 &= \frac{(y_3 - y_1)(x - x_3) + (x_1 - x_3)(y - y_3)}{(y_2 - y_3)(x_1 - x_3) + (x_3 - x_2)(y_1 - y_3)}, \\ b_3 &= 1 - b_1 - b_2 \cdot \end{aligned} \tag{7.24}$$

Remember, these equations give us the barycentric coordinates of the point x, y on the triangle, and we can verify this by seeing that $b_1 + b_2 + b_3 = 1$ and $0 \le b_1, b_2, b_3 \le 1$.

Once we have computed the barycentric coordinates of any point **p** relative to the triangle ABC, we can determine its relative position by investigating the signs of the coordinates. Figure 7.6 shows all possible regions in the plane.

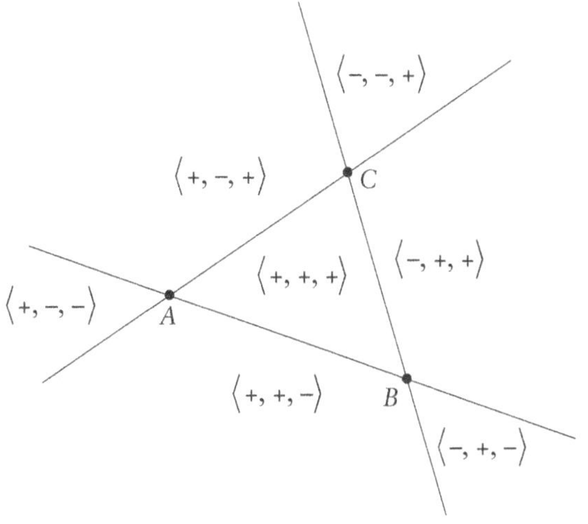

Figure 7.6 Regions in the triangle space indicating the relative position of a point

Usually, geometric models are converted to a boundary representation based on a collection of polygons that form a polyhedron. The process for doing this is called polygonization of the model. More often than not, the polygons are triangles. The three vertices of a triangle define a plane. In ray tracing for rendering a model, the intersection points of rays that occur within the model's triangle-defined planes can convey information about phenomena at the vertex points via linear interpolation using barycentric coordinates. Thus color, shading, and texture are computed for each intersect point, which in turn determine pixel values for the final computer-generated image.

7.5 Relative to a Tetrahedron

The last example of barycentric coordinates on a simplex is for the general convex tetrahedron in three-dimensional Euclidean space ("general" meaning not necessarily a regular tetrahedron). Figure 7.7 illustrates this.

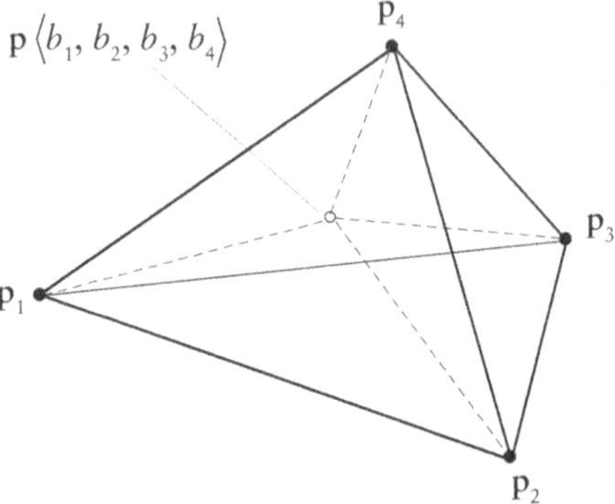

Figure 7.7 Barycentric coordinates of a tetrahedron

Given its barycentric coordinates, the coordinates of any point in the tetrahedron are

$$\begin{aligned} x &= b_1x_1 + b_2x_2 + b_3x_3 + b_4x_4, \\ y &= b_1y_1 + b_2y_2 + b_3y_3 + b_4y_4, \\ z &= b_1z_1 + b_2z_2 + b_3z_3 + b_4z_4, \end{aligned} \tag{7.25}$$

with

$$b_1 + b_2 + b_3 + b_4 = 1, \tag{7.26}$$

or

$$b_4 = 1 - b_1 - b_2 - b_3. \tag{7.27}$$

Conversely, given the point $\mathbf{p}(x, y)$ in the tetrahedron, we fall back on some matrix math: $\mathbf{AB} = \mathbf{P}$ and solving $\mathbf{B} = \mathbf{A}^{-1}\mathbf{P}$, where

$$\begin{bmatrix} b_1 \\ b_2 \\ b_3 \end{bmatrix} = \begin{bmatrix} (x_1 - x_4) & (x_2 - x_4) & (x_3 - x_4) \\ (y_1 - y_4) & (y_2 - y_4) & (y_3 - y_4) \\ (z_1 - z_4) & (z_2 - y_4) & (z_3 - z_4) \end{bmatrix}^{-1} \begin{bmatrix} (x - x_4) \\ (y - y_4) \\ (z - z_4) \end{bmatrix}. \tag{7.28}$$

Inverting the 3×3 matrix is the last step (not done here). This process applies to simplexes in higher dimensions, too. But we will stop at the tetrahedron.

7.6 Generalized Barycentric Coordinates

Now you might ask: Can we apply barycentric coordinates to other polygons and polyhedra? Yes; but we call them generalized barycentric coordinates. How are they different? The mathematical definition is the same for each system. But simplexes, the triangle and tetrahedron for example, have n vertices defined in a space of $n - 1$ dimensions. A square or quadrilateral has four vertices and is embedded in a space of two dimensions. A cube or other polyhedron with eight vertices is embedded in a space of three dimensions. A simplex with n vertices requires a space of at least $n - 1$ dimensions. A polytope with n vertices is embedded in a space of a lower dimension than $n - 1$, meaning that there are not enough constraints to produce a unique solution for the coordinates. The set of equations is underdetermined; there are not enough independent equations available to uniquely determine the values of the barycentric coordinates of any given point. Research continues in this area, and there are ways to work around this problem; their discussion would take us beyond the scope of this textbook.

8 Translation and Rotation

This chapter introduces translation and rotation, transformations that are indispensable to geometric and 3D modeling. Translation and rotation are rigid-body transformations because they do not change the size or shape of the geometric object on which they operate, only its position and orientation. The chapter ends with a discussion of homogeneous coordinates and how they facilitate combining translation and rotation transformations into a single matrix.

8.1 Translation

Translation is a rigid-body transformation, and rotation is another. Rigid-body transformations preserve distance between all points and angles between all straight lines of the transformed geometric object. This simply means that the object's size and shape don't change when the object changes position or orientation.

Let's start by looking at the simple linear equations that describe the translation of a point (x, y, z) to a new location (x', y', z'),

$$\begin{aligned} x' &= x + t_x, \\ y' &= y + t_y, \\ z' &= z + t_z, \end{aligned} \tag{8.1}$$

where t_x, t_y, and t_z are the displacements in each of the three principal directions (Figure 8.1) and where the prime superscript indicates coordinates of the transformed point.

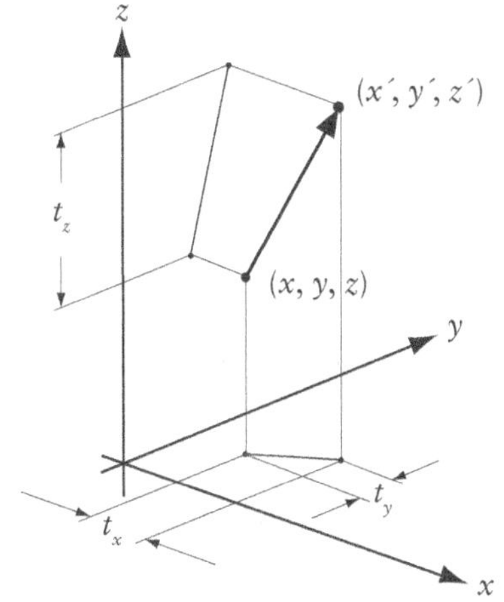

Figure 8.1 Illustration of Equation 8.1

Vectors offer a concise geometric description of a translation. If we define a point by the vector **p** and define an arbitrary translation by another vector **t**, then we express the translation of **p** by **t** (Figure 8.2) as the sum of the two vectors:

$$\mathbf{p}' = \mathbf{p} + \mathbf{t}. \tag{8.2}$$

In terms of matrices, we rewrite this to obtain

$$\mathbf{P}' = \mathbf{P} + \mathbf{T}, \tag{8.3}$$

where **P′**, **P**, and **T** are column matrices containing the vector components, as follows:

$$\begin{bmatrix} p'_x \\ p'_y \\ p'_z \end{bmatrix} = \begin{bmatrix} p_x \\ p_y \\ p_z \end{bmatrix} + \begin{bmatrix} t_x \\ t_y \\ t_z \end{bmatrix}, \tag{8.4}$$

which is just another way of writing Equation 8.1.

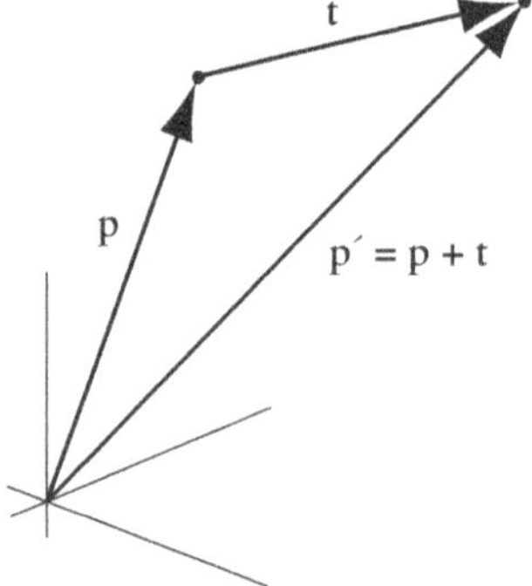

Figure 8.2 Vectors and translation of a point

Although it is tempting to think of this translation as a motion along the line of **t**, from the starting point to the ending point, it is not. There are no intermediate positions. There is no path. This also applies to rotations and other transformations that we may be tempted to see as motions. However, that said, to animate the motion of an object traveling in a straight line from A to B, many intermediate translations may be necessary. Just how many depends on how fast the object is moving, if it is moving at a uniform speed, and if the transformations are performed frequently enough to produce the illusion or perception of a smooth motion.

Here is an example of the translation of a two-dimensional figure in the xy plane: In Figure 8.3 we see a rectangle, defined by its four corner points, translated to a new position. Notice that its size, shape, and orientation are unaffected by this. So we have $\mathbf{p}'_1 = \mathbf{p}_1 + \mathbf{t}$, and similarly for the other corner points. The rectangle is reconstructed at the new location by joining the points with line segments in the same way as the original. This is nothing more than a four-sided plane polygon; $\mathbf{p}_1$, $\mathbf{p}_2$, $\mathbf{p}_3$, and $\mathbf{p}_4$ are the vertices, and $(\mathbf{p}_1 - \mathbf{p}_2)$, $(\mathbf{p}_2 - \mathbf{p}_3)$, $(\mathbf{p}_3 - \mathbf{p}_4)$, and $(\mathbf{p}_4 - \mathbf{p}_1)$ are its edges . . . similarly for its translated image.

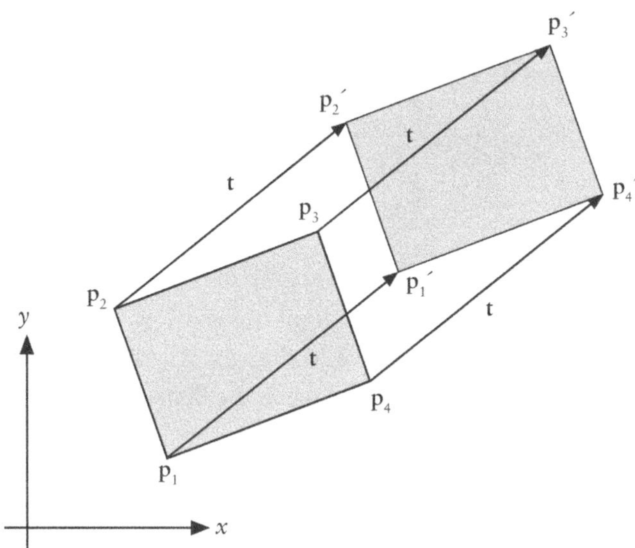

Figure 8.3 Translation of a rectangle

Here is how to use vectors to translate the straight line *L*, defined by the vector equation $\mathbf{p} = \mathbf{p}_0 + u\mathbf{a}$. The vector $\mathbf{a}$ defines the direction of the line. Because each point on the line must be translated the same direction and distance, we have (Figure 8.4)

$$\mathbf{p}' = \mathbf{p}_0 + u\mathbf{a} + \mathbf{t}. \tag{8.5}$$

Note that the translation does not change the direction of the line, so *L* and *L′* are parallel.

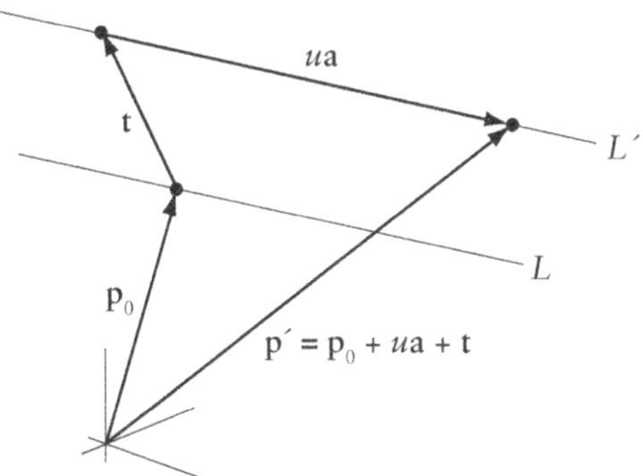

Figure 8.4 Vectors and translation of a line

What happens if we simply want to change the placement of the coordinate system (without rotation) with respect to a point or set of points? Here is a vector representation of this problem and its solution (Figure 8.5):

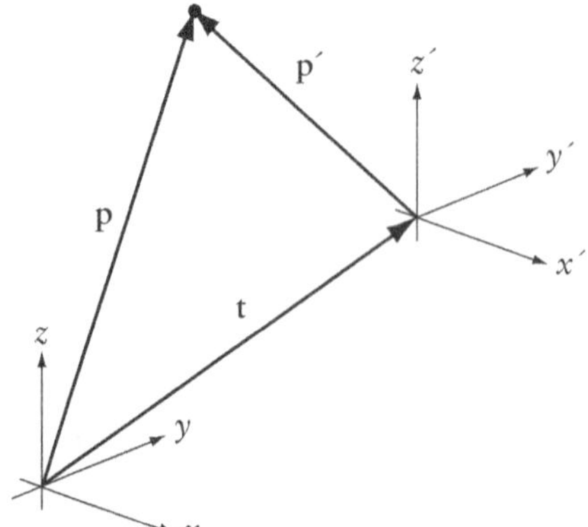

Figure 8.5 Coordinate system translation

Given a point represented by the vector **p** in the x, y, z coordinate system, find the vector representation of this point in the x', y', z' system, where the x', y', z' system is displaced relative to the x, y, z system by a translation **t**. The solution is simple enough, as you see in the figure. We find that

$$\mathbf{p}' = \mathbf{p} - \mathbf{t}. \tag{8.6}$$

In matrix form

$$\mathbf{P}' = \mathbf{P} - \mathbf{T}, \tag{8.7}$$

where **P′**, **P**, and **T** are column matrices containing vector components, as follows:

$$\mathbf{P}' = \begin{bmatrix} x' \\ y' \\ z' \end{bmatrix}, \quad \mathbf{P} = \begin{bmatrix} x \\ y \\ z \end{bmatrix}, \quad \mathbf{T} = \begin{bmatrix} t_x \\ t_y \\ t_z \end{bmatrix}. \tag{8.8}$$

Be careful of the algebraic sign of the matrix **T** and its elements, which here are the components of the translation vector. We could write this as

$$\mathbf{P}' = \mathbf{P} + \mathbf{T} \tag{8.9}$$

if we adjust the elements of **T** as follows:

$$\mathbf{T} = \begin{bmatrix} -t_x \\ -t_y \\ -t_z \end{bmatrix}. \tag{8.10}$$

We see that a coordinate system translation is equivalent to translating an object the same distance but in the opposite direction, that is, translating the object **–t**. In Cartesian coordinates this looks like

$$\begin{aligned} x' &= x - t_x, \\ y' &= y - t_y, \\ z' &= z - t_z. \end{aligned} \tag{8.11}$$

A succession of translations $\mathbf{t}_1,\ \mathbf{t}_2,\ \mathbf{t}_3,\ \ldots$ of a point $\mathbf{p}$ produces the point $\mathbf{p}'$, where

$$\mathbf{p}' = \left(\left(\left((\mathbf{p}+\mathbf{t}_1)+\mathbf{t}_2\right)+\mathbf{t}_3\right)+\ldots\right), \tag{8.12}$$

or more simply as

$$\mathbf{p}' = \mathbf{p}+\mathbf{t}_1+\mathbf{t}_2+\mathbf{t}_3+\ldots. \tag{8.13}$$

As Figure 8.6 shows, we can sum the individual translations to produce a single equivalent translation. Thus, if $\mathbf{t}=\mathbf{t}_1+\mathbf{t}_2+\mathbf{t}_3$ then $\mathbf{p}'=\mathbf{p}+\mathbf{t}$.

The order in which we compute the sum of two or more translations is not important unless they represent key vectors in an animation sequence. In the above example and in the figure, we see that

$$\begin{aligned}\mathbf{p}' &= \mathbf{p}+\mathbf{t}_1+\mathbf{t}_2+\mathbf{t}_3\\ &= \mathbf{p}+\mathbf{t}_1+\mathbf{t}_3+\mathbf{b}_2\\ &= \mathbf{p}+\mathbf{t}_2+\mathbf{t}_1+\mathbf{t}_3,\end{aligned} \tag{8.14}$$

and so forth. From this we conclude that a succession of transformations is commutative.

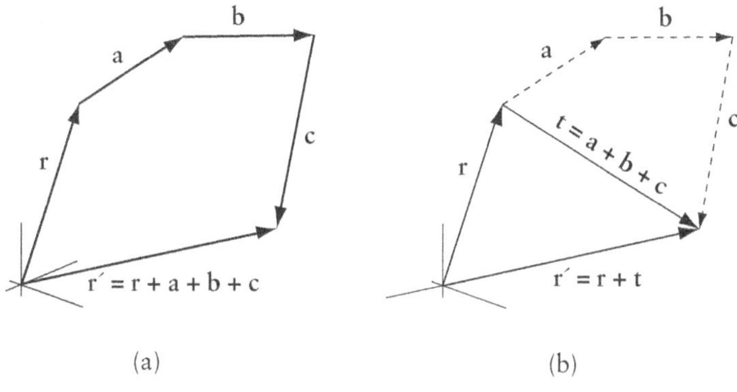

Figure 8.6 A succession of translations

Translating points, lines, and planes is easy to explain and understand. But what about curves and surfaces, or 3D solids for that matter? Well, as it happens, most geometric objects are defined by a set of points, usually known as control points. A Bézier curve is an example. We define a cubic Bézier curve with a set of four control points. To translate this kind of curve, we simply translate each of its four control points by **t** (Figure 8.7), and that's it.

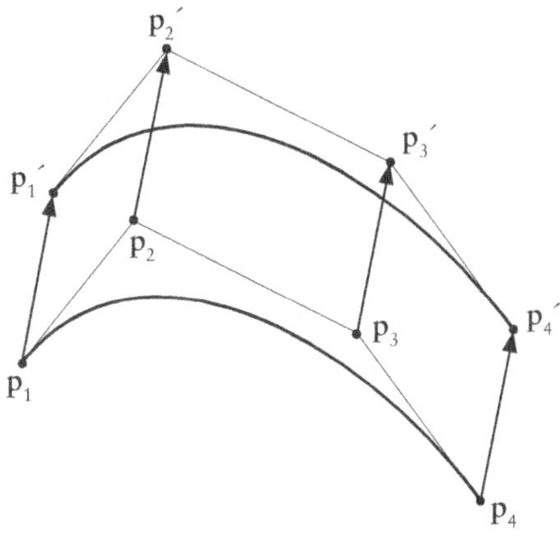

Figure 8.7 Translation of a Bézier curve

For a Hermite curve, which is defined by its two endpoints and tangent vectors, simply translate the endpoints. The tangent vectors do not change direction or magnitude under translation transformations.

8.2 Rotation in the Plane

There are many ways to impose a rotation transformation on a geometric model. There are rotations in the plane and 3D space relative to the origin or coordinate axes. There are rotations about an arbitrary axis in space. There are equivalent rotations revealed by eigenvectors, and there are rotations executed via quaternions. These are topics covered in this and the following sections.

A rotation about the origin in the xy plane is the simplest kind of rotation. (It is so simple that it obscures many characteristics of rotation in three dimensions.) We will use a right-hand coordinate system and the convention that positive rotations are counterclockwise.

The coordinates of the transformed point $\mathbf{p}'$ are functions of the coordinates of $\mathbf{p}$ and the rotation angle ϕ. The derivation of these functions is easy, and Figure 8.8 shows some of the details.

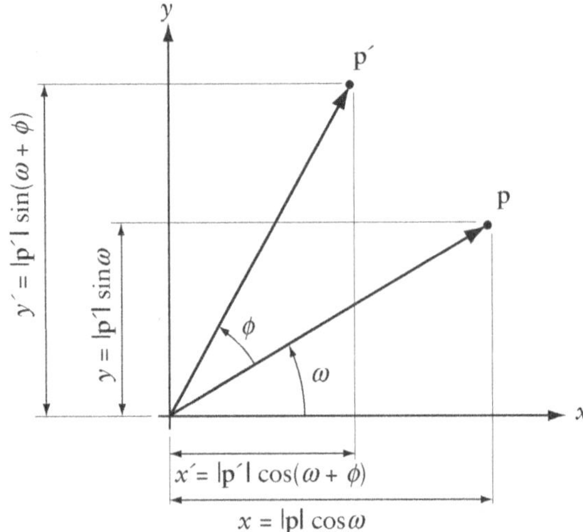

Figure 8.8 Rotation in the xy plane

In the figure, we see that

$$x' = |\mathbf{p}'| \cos(\omega + \phi),$$
$$y' = |\mathbf{p}'| \sin(\omega + \phi), \tag{8.15}$$

and

$$\cos\omega = \frac{x}{|\mathbf{p}|}, \quad \sin\omega = \frac{y}{|\mathbf{p}|}, \tag{8.16}$$

where ω is the angle between $\mathbf{p}'$ and the x axis.

Because this is a rigid-body transformation, it preserves the distance between points (in this case the origin and point **p**), and we see that $|\mathbf{p}'| = |\mathbf{p}|$. These two points are on the circumference of a circle whose center is at the origin. From elementary trigonometric identities, we know that

$$\begin{aligned} \cos(\omega+\phi) &= \cos\omega\cos\phi - \sin\omega\sin\phi, \\ \sin(\omega+\phi) &= \sin\omega\cos\phi + \cos\omega\sin\phi. \end{aligned} \tag{8.17}$$

With some appropriate algebra we obtain

$$\begin{aligned} x' &= x\cos\phi - y\sin\phi, \\ y' &= x\sin\phi + y\cos\phi. \end{aligned} \tag{8.18}$$

This is the set of rotation transformation equations we are after. It is important to note that these equations are immediately useful to rotations in three dimensions . . . as we shall soon see.

Writing Equation 8.18 in matrix form produces

$$\begin{bmatrix} x' \\ y' \end{bmatrix} = \begin{bmatrix} \cos\phi & -\sin\phi \\ \sin\phi & \cos\phi \end{bmatrix} \begin{bmatrix} x \\ y \end{bmatrix}, \tag{8.19}$$

or

$$\mathbf{P}' = \mathbf{R}_\phi \mathbf{P}, \tag{8.20}$$

where $\mathbf{R}_\phi$ is the rotation matrix

$$\mathbf{R}_\phi = \begin{bmatrix} \cos\phi & -\sin\phi \\ \sin\phi & \cos\phi \end{bmatrix}, \tag{8.21}$$

and **P** is a column matrix containing the vector components of the initial point, and **P′** contains the components of the transformed point. Here we see that the rotation transformation matrix premultiplies the point coordinate matrix. Compare this with the translation as presented above (Equation 8.9), where the point and translation matrices are added. This difference is a problem we will look at later, and its resolution.

Note that the determinant $|\mathbf{R}_\phi| = 1$. This means that $\mathbf{R}_\phi$ is an orthogonal matrix and conforms to the restriction on the coefficients for a rigid-body transformation.

To reverse it, we apply the rotation $-\phi$ and obtain

$$\mathbf{R}_{-\phi} = \begin{bmatrix} \cos\phi & \sin\phi \\ -\sin\phi & \cos\phi \end{bmatrix}. \tag{8.22}$$

Using matrix algebra, we find that

$$\mathbf{R}_{-\phi}\mathbf{R}_{\phi} = \begin{bmatrix} 1 & 0 \\ 0 & 1 \end{bmatrix}, \tag{8.23}$$

or

$$\mathbf{R}_{-\phi}\mathbf{R}_{\phi} = \mathbf{I}. \tag{8.24}$$

This means that

$$\mathbf{R}_{-\phi} = \mathbf{R}_{\phi}^{-1}. \tag{8.25}$$

We can compute the product of two successive rotations about the origin in the xy plane as follows: First, staying with the matrix form, we rotate **P** by ϕ_1 to obtain **P′**:

$$\mathbf{P}' = \mathbf{R}_{\phi 1}\mathbf{P}. \tag{8.26}$$

Second, we rotate **P′** by ϕ_2 to obtain **P″**, where

$$\mathbf{P}'' = \mathbf{R}_{\phi 2}\mathbf{P}', \tag{8.27}$$

or

$$\mathbf{P}'' = \mathbf{R}_{\phi 2}\mathbf{R}_{\phi 1}\mathbf{P}. \tag{8.28}$$

The product of the two rotation matrices yields

$$\mathbf{R}_{\phi 2}\mathbf{R}_{\phi 1} = \begin{bmatrix} \cos\phi_1\cos\phi_2 - \sin\phi_1\sin\phi_2 & -\sin\phi_1\cos\phi_2 - \cos\phi_1\sin\phi_2 \\ \cos\phi_1\sin\phi_2 + \sin\phi_1\cos\phi_2 & -\sin\phi_1\sin\phi_2 + \cos\phi_1\cos\phi_2 \end{bmatrix}. \tag{8.29}$$

Again, we use common trigonometric identities to obtain

$$\mathbf{R}_{\phi 2}\mathbf{R}_{\phi 1} = \begin{bmatrix} \cos(\phi_1+\phi_2) & -\sin(\phi_1+\phi_2) \\ \sin(\phi_1+\phi_2) & \cos(\phi_1+\phi_2) \end{bmatrix}. \tag{8.30}$$

This tells us that for two successive rotations in the plane about the origin, we merely add the two rotation angles ϕ_1 and ϕ_2 and use the resulting sum in the rotation matrix. Of course, this extends to the product of n successive rotations, so that

$$\mathbf{R} = \begin{bmatrix} \cos(\phi_1+\phi_2+\ldots+\phi_n) & -\sin(\phi_1+\phi_2+\ldots+\phi_n) \\ \sin(\phi_1+\phi_2+\ldots+\phi_n) & \cos(\phi_1+\phi_2+\ldots+\phi_n) \end{bmatrix}. \tag{8.31}$$

To execute a rotation in the plane about some point other than the origin, we let $\mathbf{p}_c$ and α define the center and angle of rotation, respectively. and then proceed using the following 3 steps (Figure 8.9):

1. Translate $\mathbf{p}_c$ to the origin. Translate all points defining a figure in the same direction and distance: $\mathbf{p} - \mathbf{p}_c$.

2. Rotate the results about the origin: $\mathbf{R}_\alpha (\mathbf{P} - \mathbf{P}_c)$. (In Figure 8.9 $\alpha = 90°$.)
3. Reverse Step 1, returning the center of rotation to $\mathbf{p}_c$ and thus producing

$$\mathbf{P}' = \mathbf{R}_\alpha (\mathbf{P} - \mathbf{P}_c) + \mathbf{P}_c. \tag{8.32}$$

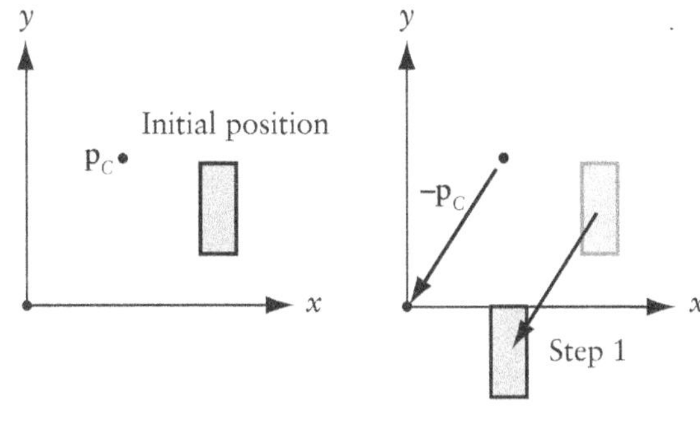

Figure 8.9 Rotation about an arbitrary point

Rotating the coordinate system about the origin and through an angle ϕ_c is equivalent to an oppositely directed rotation of points in the original system (Figure 8.10). This means that $\phi = -\phi_c$, and the rotation matrix becomes

$$\mathbf{R}_{-\phi_c} = \begin{bmatrix} \cos\phi_c & \sin\phi_c \\ -\sin\phi_c & \cos\phi_c \end{bmatrix}. \tag{8.33}$$

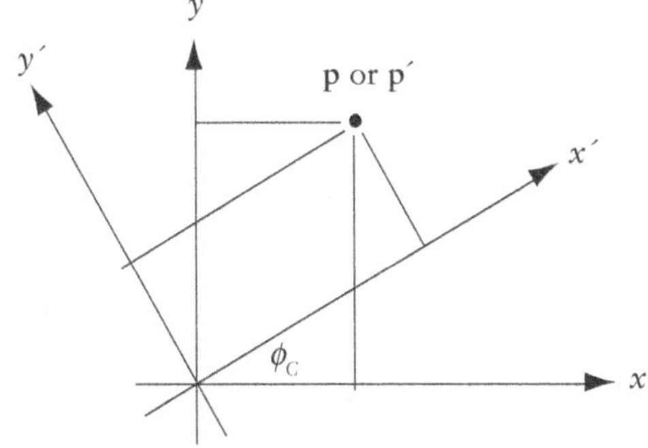

Figure 8.10 Rotating the coordinate system about the origin and through an angle ϕ_c

8.3 Rotation in Space

The simplest rotations in space are those about one or more of the principal axes, or about some axis through the origin. Other rotations are variations of these two, where the first step is to bring the axis of rotation to the origin or in alignment with one of the principal axes, perform the rotation, and finally reverse the first step. Again, we apply the right-hand rule to define the algebraic sign of a rotation, where the thumb points in the axis's positive direction.

Let's start by investigating rotations about the principal axes. If we look toward the origin from a point on the z axis, a positive rotation is counterclockwise; similarly for the x and y axes (Figure 8.11). Rotation about the z axis produces a result identical to a rotation about the origin in the xy plane. However, we must account for the z coordinate. We do this as follows:

$$\begin{aligned} x' &= x\cos\phi - y\sin\phi, \\ y' &= x\sin\phi + y\cos\phi, \\ z' &= z. \end{aligned} \tag{8.34}$$

For this rotation the z coordinates are not changed.

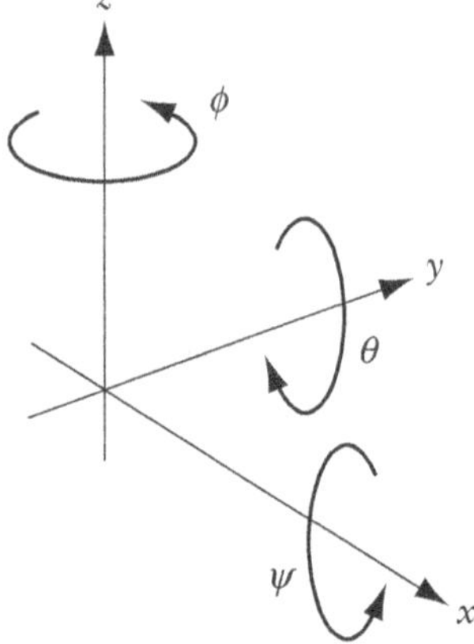

Figure 8.11 The right-hand rule defines the algebraic sign of a rotation

The rotation matrix for this transformation is

$$\mathbf{R}_\phi = \begin{bmatrix} \cos\phi & -\sin\phi & 0 \\ \sin\phi & \cos\phi & 0 \\ 0 & 0 & 1 \end{bmatrix}. \tag{8.35}$$

For rotation about the x or y axis, the matrices are

$$\mathbf{R}_\varphi = \begin{bmatrix} 1 & 0 & 0 \\ 0 & \cos\varphi & -\sin\varphi \\ 0 & \sin\varphi & \cos\varphi \end{bmatrix} \tag{8.36}$$

and

$$\mathbf{R}_\theta = \begin{bmatrix} \cos\theta & 0 & \sin\theta \\ 0 & 1 & 0 \\ -\sin\theta & 0 & \cos\theta \end{bmatrix}. \tag{8.37}$$

Remember that $|\mathbf{R}_\phi| = |\mathbf{R}_\varphi| = |\mathbf{R}_\theta| = 1$.

We can describe the rotation of a point or a point-defined figure in space as a product of successive rotations about each of the principal axes. First, we must establish the sequence of rotations necessary to achieve the desired effect, because the order in which we execute rotations in space is important. Different orders produce very different results. Here is one sequence:

First, perform the rotation about the z axis, $\mathbf{R}_\phi$.
Next, perform the rotation about the y axis, $\mathbf{R}_\theta$.
Finally, perform the rotation about the x axis, $\mathbf{R}_\varphi$.

The matrix equation for this sequence of rotations is

$$\mathbf{P}' = \mathbf{R}_\varphi \mathbf{R}_\theta \mathbf{R}_\phi \mathbf{P}. \tag{8.38}$$

If we let $\mathbf{R}_{\varphi\theta\phi} = \mathbf{R}_\varphi \mathbf{R}_\theta \mathbf{R}_\phi$, then

$$\mathbf{P}' = \mathbf{R}_{\varphi\theta\phi} \mathbf{P}. \tag{8.39}$$

Multiplying the rotation matrices produces Equation 8.40, which follows:

$$\mathbf{R}_{\varphi\theta\phi} = \begin{bmatrix} \cos\theta\cos\phi & -\cos\theta\sin\phi & \sin\theta \\ \cos\varphi\sin\phi + \sin\varphi\sin\theta\cos\phi & \cos\varphi\cos\phi - \sin\varphi\sin\theta\sin\phi & -\sin\varphi\cos\theta \\ \sin\varphi\sin\phi - \cos\varphi\sin\theta\cos\phi & \sin\varphi\cos\phi + \cos\varphi\sin\theta\sin\phi & \cos\varphi\cos\theta \end{bmatrix}. \tag{8.40}$$

We have already seen that $|\mathbf{R}_\varphi| = |\mathbf{R}_\theta| = |\mathbf{R}_\phi| = 1$, and from matrix algebra, we can show that

$$|\mathbf{R}_{\varphi\theta\varphi}| = |\mathbf{R}_\varphi||\mathbf{R}_\theta||\mathbf{R}_\phi| = 1. \tag{8.41}$$

Reminder: Order is important. For example,

$$\mathbf{R}_\varphi \mathbf{R}_\theta \mathbf{R}_\phi \neq \mathbf{R}_\theta \mathbf{R}_\varphi \mathbf{R}_\phi, \tag{8.42}$$

as is demonstrated here by Equation 8.43, which follows:

$$\mathbf{R}_{\theta\varphi\phi} = \begin{bmatrix} \cos\theta\cos\phi + \sin\theta\sin\varphi & -\cos\theta\sin\phi + \sin\theta\cos\phi\sin\varphi & -\sin\theta\sin\varphi \\ \cos\varphi\sin\phi & \cos\varphi\cos\phi & -\sin\varphi \\ -\sin\theta\cos\phi - \cos\theta\sin\varphi & \sin\theta\cos\phi + \cos\theta\sin\phi\sin\varphi & \cos\theta\cos\varphi \end{bmatrix}. \tag{8.43}$$

8.4 Rotation about an Arbitrary Axis

The most general rotation transformation of points and point-defined geometric models is one through some angle α about some arbitrary axis **a** in space that passes through the origin of the coordinate system (Figure 8.12). The development of this transformation that follows demonstrates the power of vectors and how we can use vector mathematics to solve a complex transformation problem.

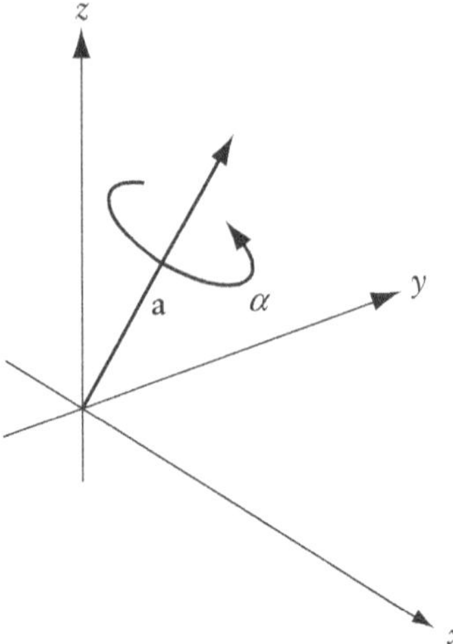

Figure 8.12 Rotation through some angle α about some arbitrary axis **a**

We begin with a sketch of what we know (Figure 8.13). First, we have an arbitrary axis in space through the origin about which we will rotate a point or points. We define the axis by the unit vector $\hat{\mathbf{a}}$. The vector **p** represents the point to be rotated through an angle α. The vector $\mathbf{p}'$ represents the rotated point. Notice that the arc defined by the sweep of α lies in a plane perpendicular to the axis of rotation. Also, note that the lengths of **p** and $\mathbf{p}'$ are equal. We will see that the two vectors **p** and $\hat{\mathbf{a}}$ and the angle α are mathematically sufficient to determine the rotated position $\mathbf{p}'$. This section demonstrates how to compute $\mathbf{p}'$ using just these three variables.

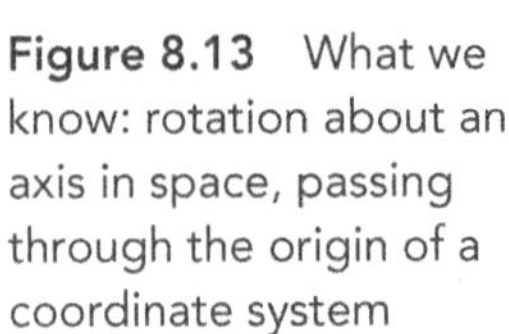

Figure 8.13 What we know: rotation about an axis in space, passing through the origin of a coordinate system

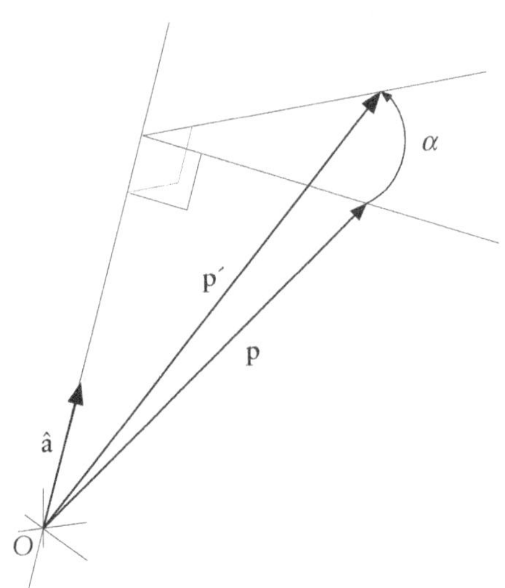

First, we use vectors to develop some supporting geometry (Figure 8.14). Next, we will decompose **p** into two components: its projection onto the rotation axis line, producing a vector that runs from O (the origin) to a point we label 1, and the vector **b** that runs from point 1 to point 2. Clearly the sum of these two vectors produces **p**. Next, we create the vector **c** from point 1 to point 3 that is perpendicular to both $\hat{\mathbf{a}}$ and **b**, such that $|\mathbf{c}| = |\mathbf{b}|$. This last construction allows us to create a useful coordinate system with $\hat{\mathbf{b}}$ and $\hat{\mathbf{c}}$ as basis vectors. Because the rotation axis vector $\hat{\mathbf{a}}$ is a unit vector, we use the property that the vector product of two perpendicular unit vectors is another unit vector, so that

$$\hat{\mathbf{c}} = \hat{\mathbf{a}} \times \hat{\mathbf{b}}. \tag{8.44}$$

We also observe that

$$|\mathbf{b}| = |\mathbf{c}| = |\mathbf{b}'|, \tag{8.45}$$

and that we can express the components of $\mathbf{b}'$ in the $\hat{\mathbf{b}}$, $\hat{\mathbf{c}}$ basis coordinate system as functions of **b**, **c**, and α.

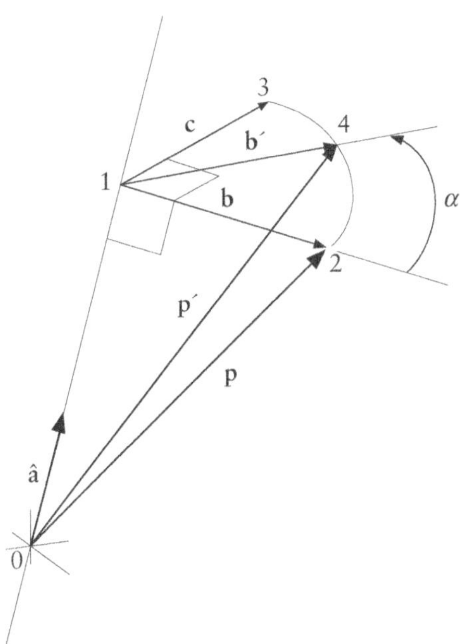

Figure 8.14 Toward a solution

Now consider this: This transformation does not change the length of **p**; in other words, $|\mathbf{p}'| = |\mathbf{p}|$. This also means that the component of $\mathbf{p}'$ that is its projection onto the axis of rotation $\hat{\mathbf{a}}$ is identical to the projection of **p** onto that axis. Any rotation of **p** or $\mathbf{p}'$ about **a** does not affect this component. We will make good use of this in the next step (Figure 8.15).

The vector projection of **p** onto $\hat{\mathbf{a}}$ is $(\mathbf{p} \bullet \hat{\mathbf{a}})\hat{\mathbf{a}}$. This produces one of the components we are interested in. Remember that **a** is a unit vector.

Next, we find the component **b** that is orthogonal to the rotation axis **â**. This is a simple matter of vector arithmetic:

$$\mathbf{b} = \mathbf{p} - (\mathbf{p} \bullet \hat{\mathbf{a}})\hat{\mathbf{a}}. \tag{8.46}$$

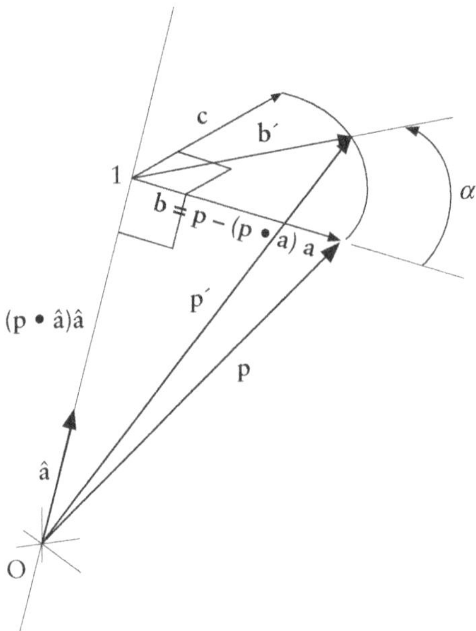

Figure 8.15 Useful components of **p**

The next piece of the puzzle is to find **b′**. The length of **b′** is the same as that of **b**. This is true because **b** is undergoing a rigid-body rotational displacement into **b′** through the angle α. We apply some trigonometry and vector properties to compute **b′**. Once we do this, we simply add **b′** and $(\mathbf{p} \bullet \hat{\mathbf{a}})\hat{\mathbf{a}}$ to produce **p′**. Let's do another figure, one showing the situation in the plane defined by **b** and **c** (Figure 8.16).

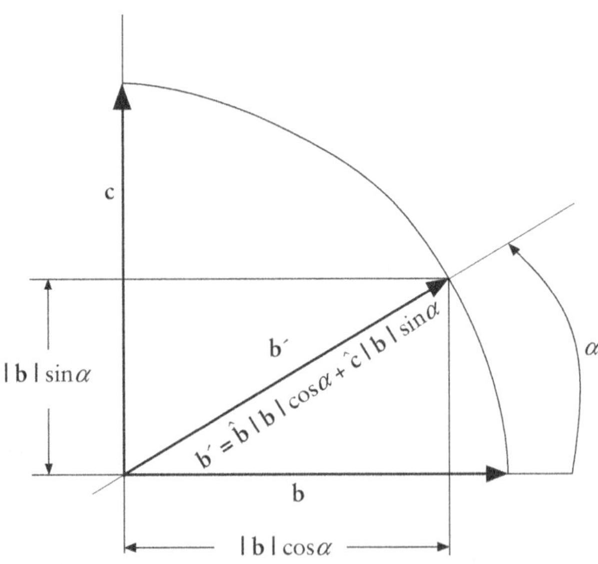

Figure 8.16 Geometry in the plane of **b** and **c**

The components of $\mathbf{b}'$ in the $\hat{\mathbf{b}}$ and $\hat{\mathbf{c}}$ basis system are $|\mathbf{b}|\cos\alpha$ and $|\mathbf{b}|\sin\alpha$, respectively. We use these coordinates to construct $\mathbf{b}'$, which we can now write as

$$\mathbf{b}' = (|\mathbf{b}|\cos\alpha)\hat{\mathbf{b}} + (|\mathbf{b}|\sin\alpha)\hat{\mathbf{c}}. \tag{8.47}$$

And finally, we add $\mathbf{b}'$ to $(\mathbf{p}\bullet\hat{\mathbf{a}})\hat{\mathbf{a}}$ to find

$$\mathbf{p}' = (\mathbf{p}\bullet\hat{\mathbf{a}})\hat{\mathbf{a}} + (|\mathbf{b}|\cos\alpha)\hat{\mathbf{b}} + (|\mathbf{b}|\sin\alpha)\hat{\mathbf{c}}. \tag{8.48}$$

Here is a figure, similar to many of those above, which sums up the results of this derivation. Figure 8.17 shows the initial vector $\mathbf{p}$ and the result of rotating it, $\mathbf{p}'$, including their components.

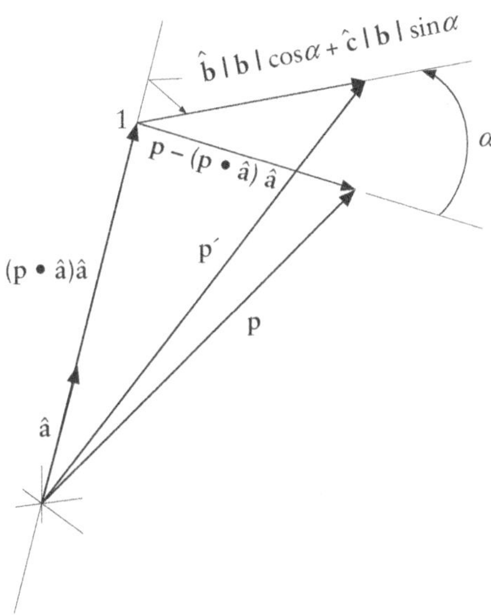

Figure 8.17 The components of **p** and **p′**

Now we do some simple algebraic substitutions to state this equation solely in terms of the original principal vectors $\mathbf{p}$ and $\hat{\mathbf{a}}$ and the rotation angle α. Here is the result:

$$\mathbf{p}' = \mathbf{p}\cos\alpha + (\hat{\mathbf{a}}\times\mathbf{p})\sin\alpha + \hat{\mathbf{a}}(\hat{\mathbf{a}}\bullet\mathbf{p})(1-\cos\alpha). \tag{8.49}$$

This equation is Rodrigues's formula, named after Olinde Rodrigues, who derived its mathematical foundations and included this in his doctoral dissertation (1815).

The Rodrigues rotation formula has a matrix form. The details of its derivation are not given here but are straightforward to work out. Here is the formula:

$$\mathbf{R} = \mathbf{I} + (\sin\alpha)\mathbf{A} + (1-\cos\alpha)\mathbf{A}^2, \tag{8.50}$$

where $\mathbf{R}$ is the 3×3 rotation matrix, $\mathbf{I}$ is the 3×3 identity matrix, α is the rotation angle, and $\mathbf{A}$ is the cross-product matrix of the unit vector $\hat{\mathbf{a}}$ defining the axis of rotation. This is what the matrix $\mathbf{A}$ looks like:

$$\mathbf{A} = \begin{bmatrix} 0 & -a_z & a_y \\ a_z & 0 & -a_x \\ -a_y & a_x & 0 \end{bmatrix}. \tag{8.51}$$

So the product $\mathbf{p}'$ of a rotation of any vector $\mathbf{p}$ about $\hat{\mathbf{a}}$ through the angle α in matrix form looks like

$$\mathbf{P}' = \mathbf{RP}. \tag{8.52}$$

8.5 Eigenvectors and Equivalent Rotations

Eigenvalues and eigenvectors prove to be useful to our understanding of rotations in space. We know that a rotation in the plane fixes one point. A rotation or a series of rotations in space (say, about axes through the origin) fixes a line, the equivalent axis of rotation. We represent this axis by some unit vector $\hat{\mathbf{a}}$. Given a rotation matrix $\mathbf{R}$ describing some arbitrary rotation, perhaps the net result of successive rotations about the principal axes, it follows that

$$\mathbf{R}\hat{\mathbf{a}} = \lambda\hat{\mathbf{a}}. \tag{8.53}$$

In other words, $\hat{\mathbf{a}}$ must be an eigenvector of $\mathbf{R}$, so we apply the mathematics of eigenvalues and eigenvectors to find $\hat{\mathbf{a}}$ given $\mathbf{R}$.

Now consider Euler's theorem, which states that the general displacement transformation of a rigid body with one fixed point is a rotation about some axis. If the fixed point is the origin, then the displacement produces no translation of the body system, and only rotational motion is present. Furthermore, we know that it is a characteristic of any rotation in space that one line is fixed, the axis of rotation. Any vector along the axis of rotation must have the same components in both the initial and final position of the body, where the orthogonality property preserves the magnitude of the vector. We can demonstrate this by showing that there exists a vector $\mathbf{a}$ having the same components in both systems. This is the eigenvalue problem, and the solution is the eigenvector(s) of $\mathbf{R}$. Now we can restate Euler's theorem this way: The real orthogonal matrix specifying the physical motion of a rigid body with one point fixed has an eigenvalue equal to 1.

Given that an eigenvector of $\mathbf{R}$ for the eigenvalue $\lambda = 1$ is in the direction of the axis, we then compute it as any column vector of the matrix of cofactors of $\mathbf{R} - \mathbf{I}$. We compute the angle of rotation α from the relationship:

$$\mathrm{tr}\mathbf{R} = 1 + 2\cos\alpha. \tag{8.54}$$

The sign of α is the same as the sign of the evaluated 3x3 determinant $\det(\mathbf{p},\ \mathbf{Rp},\ \hat{\mathbf{a}})$ formed from the column vectors $\mathbf{p}$, $\mathbf{Rp}$, and $\hat{\mathbf{a}}$, where $\mathbf{p}$ is any vector not parallel to $\hat{\mathbf{a}}$. The right-hand rule applies. Notice that the sign of α changes if we replace $\hat{\mathbf{a}}$ with $-\hat{\mathbf{a}}$. You can find a general proof of this in more advanced discussions of rotations.

We can also compute the eigenvector somewhat more efficiently as any column vector of the symmetric matrix:

$$\left(\mathbf{R}+\mathbf{R}^{\mathrm{T}}\right)-(\mathrm{tr}\mathbf{R}-1)\mathbf{I}. \tag{8.55}$$

Try to work out the proof of this assertion.

Recall (Chapter 4) that the trace tr(**A**) of a square matrix **A** is the sum of its diagonal entries:

$$\mathrm{tr}(\mathbf{A}) = a_{11} + a_{22} + \ldots + a_{nn}, \text{ or } \mathrm{tr}(\mathbf{A}) = \sum_{i=1}^{n} a_{ii}.$$

8.6 Rotation and Quaternions

Quaternion algebra offers an effective way to execute rotation transformations in certain circumstances. Let the unit vector $\hat{\mathbf{a}}$ represent the axis of rotation passing through the origin, and let α represent the angle of rotation about this axis. We'll represent this rotation by the quaternion R:

$$R = \cos(\theta/2) + \hat{\mathbf{a}}\sin(\theta/2). \tag{8.56}$$

To rotate a point $\mathbf{p}$ through θ about $\hat{\mathbf{a}}$, perform the quaternion multiplication

$$P' = RPR^{-1}, \tag{8.57}$$

where P is the quaternion representing $\mathbf{p}$ (i.e., the scalar part of P is zero, and the vector part is simply $\mathbf{p}$) and where P' is the rotated quaternion from which we extract $\mathbf{p}'$.

Although we will not formally prove the validity of the two expressions above, it is easy to show that when P represents a vector and R a rotation, then RPR^{-1} also represents a vector. Simply carry out the indicated quaternion multiplication and inspect the form of P'. Also, compare RPR^{-1} with the standard rotation product **RP**.

To eliminate the half-angle notation, we'll use the trigonometric identities

$$\begin{aligned}\cos 2\theta &= \cos^2\theta - \sin^2\theta \\ &= 1 - 2\sin^2\theta\end{aligned} \tag{8.58}$$

and

$$\sin 2\theta = 2\sin\theta\cos\theta. \tag{8.59}$$

When all the math is done, and when it is translated into a vector format, we find that

$$\mathbf{p}' = \mathbf{p}\cos\theta + (\hat{\mathbf{a}}\times\mathbf{p})\sin\theta + \hat{\mathbf{a}}(\hat{\mathbf{a}}\bullet\mathbf{p})(1-\cos\theta), \tag{8.60}$$

where θ is the angle of rotation of $\mathbf{p}$ and where $\hat{\mathbf{a}}$ is the unit vector defining the axis of rotation. Compare this equation with the Rodrigues formula, Equation 8.49. Yes, they are identical.

8.7 Combining Translation and Rotation

Rotation acts via a 3×3 matrix applied to a point or series of points that define a geometric object. It premultiplies the 3×1 point matrices. If a sequence of these transformations is imposed on a point, then matrix mathematics allows us to first multiply together the separate rotation transformations to produce a single combined transformation matrix, which then premultiplies the point matrix to produce the transformed point.

The translation transformation is different; it is accomplished with matrix addition . . . and this really messes up a sequence of transformations that includes one or more translations with rotations. It makes it impossible to construct a single composite transformation matrix.

Let's look more closely at the problem caused by translations. For example, to rotate an object about an axis not through the origin, we combine it with appropriate translations. A sequence of transformations of this type quickly becomes a nested mess of matrix additions and multiplications. We must find a way to represent rigid-body translations in a form such that we can express a mixed sequence of transformations of all types, including both rotations and translations, as a product of the matrices describing these transformations that produces a single equivalent matrix.

To do this we use homogeneous coordinates.

What?!

Okay, it's time for a review. From your algebra studies, a homogenous equation is one in which all the terms are of the same degree. Here is an example:

$$ax + by + cz = 0, \tag{8.61}$$

and here is an equation that is not homogeneous:

$$ax + by + cz + d = 0, \tag{8.62}$$

where the fourth term differs from the first three.

Let's use point coordinates as an example. Introducing an extra coordinate will make the transformation equations homogeneous. We will strip out this extra coordinate when necessary, leaving us with our ordinary Cartesian coordinates. This is how we represent a point in homogeneous coordinates in matrix form:

$$\mathbf{P}_{\mathrm{h}} = \begin{bmatrix} x \\ y \\ z \\ 1 \end{bmatrix}. \tag{8.63}$$

Right away, this allows us to express a translation transformation in matrix form as the product of the transformation matrix $\mathbf{T}_{\mathrm{h}}$ and the homogeneous point matrix $\mathbf{P}_{\mathrm{h}}$. (We will soon drop the subscript h.)

$$\mathbf{P}'_{\mathrm{h}} = \mathbf{T}_{\mathrm{h}}\mathbf{P}_{\mathrm{h}}, \tag{8.64}$$

where the homogeneous translation transformation matrix $\mathbf{T}_{\mathrm{h}}$ is

$$\mathbf{T}_{\mathrm{h}} = \begin{bmatrix} 1 & 0 & 0 & t_x \\ 0 & 1 & 0 & t_y \\ 0 & 0 & 1 & t_z \\ 0 & 0 & 0 & 1 \end{bmatrix}. \tag{8.65}$$

To make this approach even more useful (we'll see how later), we introduce yet another modification to our matrix representation of a point by generalizing it, adding a scalar multiplier h to each of the coordinates. This produces

$$\mathbf{P}_{\mathrm{h}} = \begin{bmatrix} hx \\ hy \\ hz \\ h \end{bmatrix}. \tag{8.66}$$

The Cartesian coordinates are then simply

$$x = \frac{hx}{h}, \quad y = \frac{hy}{h}, \quad z = \frac{hz}{h}. \tag{8.67}$$

Introducing the scalar term h allows us to describe certain projective transformations (not discussed here). For most other transformations, we assume $h = 1$ unless indicated otherwise.

Notice that if we use the generalized homogeneous coordinates, then the representation of a point in Cartesian space is no longer unique. For example, the homogeneous coordinates (6, 10, 2, 2) and (9, 15, 3, 3) both represent the same point (3, 5, 1) in three-dimensional Cartesian space. Furthermore, we assert that any coordinate system representing a point in three-dimensional space using four coordinates, where any scalar multiple $h\mathbf{p}$ of a given vector $\mathbf{p}$ represents the same point in 3D space, is a homogeneous coordinate system. We easily generalize this further with the following observation: The homogeneous coordinates of a point in n-dimensional space consist of $n + 1$ numbers.

Here are some examples of how it operates on single transformations. For a point in 2D space, we have

$$\mathbf{P} = \begin{bmatrix} x \\ y \\ 1 \end{bmatrix}. \tag{8.68}$$

For a point in 3D space, we have

$$\mathbf{P} = \begin{bmatrix} x \\ y \\ z \\ 1 \end{bmatrix}. \tag{8.69}$$

For rotation in the plane, the matrix transformation $\mathbf{P}' = \mathbf{RP}$ looks like this:

$$\begin{bmatrix} x' \\ y' \end{bmatrix} = \begin{bmatrix} \cos\phi & -\sin\phi \\ \sin\phi & \cos\phi \end{bmatrix} \begin{bmatrix} x \\ y \end{bmatrix}. \tag{8.70}$$

After we adjust it to accommodate homogeneous coordinates, it looks like this:

$$\begin{bmatrix} x' \\ y' \\ 1 \end{bmatrix} = \begin{bmatrix} \cos\phi & -\sin\phi & 0 \\ \sin\phi & \cos\phi & 0 \\ 0 & 0 & 1 \end{bmatrix} \begin{bmatrix} x \\ y \\ 1 \end{bmatrix}. \tag{8.71}$$

In this equation, we express the initial $\mathbf{P}$ and transformed $\mathbf{P}'$ coordinates in homogeneous coordinates, and similarly for the transformation matrix $\mathbf{R}$. When we perform the matrix multiplication, we obtain

$$\begin{aligned} x' &= x\cos\phi - y\sin\phi, \\ y' &= x\sin\phi + y\cos\phi, \\ 1 &= 1. \end{aligned} \tag{8.72}$$

We disregard the last identity, $1 = 1$, and what remains is what we would expect.

For a translation in the plane, $\mathbf{P}' = \mathbf{TP}$, we write

$$\begin{bmatrix} x' \\ y' \\ 1 \end{bmatrix} = \begin{bmatrix} 1 & 0 & t_x \\ 0 & 1 & t_y \\ 0 & 0 & 1 \end{bmatrix} \begin{bmatrix} x \\ y \\ 1 \end{bmatrix}. \tag{8.73}$$

Performing the matrix multiplication and eliminating the homogeneous coordinate, we obtain

$$\begin{aligned} x' &= x + t_x, \\ y' &= y + t_y. \end{aligned} \tag{8.74}$$

We translate points in three-dimensional space this way:

$$\begin{bmatrix} x' \\ y' \\ z' \\ 1 \end{bmatrix} = \begin{bmatrix} 1 & 0 & 0 & t_x \\ 0 & 1 & 0 & t_y \\ 0 & 0 & 1 & t_z \\ 0 & 0 & 0 & 1 \end{bmatrix} \begin{bmatrix} x \\ y \\ z \\ 1 \end{bmatrix}, \tag{8.75}$$

where t_x, t_y, and t_z are components of the vector **t** that describes the transformation $\mathbf{p}' = \mathbf{p} + \mathbf{t}$. This matrix product yields

$$\begin{bmatrix} x' \\ y' \\ z' \\ 1 \end{bmatrix} = \begin{bmatrix} x + t_x \\ y + t_y \\ z + t_z \\ 1 \end{bmatrix}, \tag{8.76}$$

from which we readily compute the transformed Cartesian coordinates:

$$\begin{aligned} x' &= x + t_x, \\ y' &= y + t_y, \\ z' &= z + t_z. \end{aligned} \tag{8.77}$$

To rotate a point around a principal axis, we use one of the three following transformation equations:

- For rotation around the z axis, we have

$$\begin{bmatrix} x' \\ y' \\ z' \\ 1 \end{bmatrix} = \begin{bmatrix} \cos\theta & -\sin\theta & 0 & 0 \\ \sin\theta & \cos\theta & 0 & 0 \\ 0 & 0 & 1 & 0 \\ 0 & 0 & 0 & 1 \end{bmatrix} \begin{bmatrix} x \\ y \\ z \\ 1 \end{bmatrix}. \tag{8.78}$$

- For rotation around the y axis, we have

$$\begin{bmatrix} x' \\ y' \\ z' \\ 1 \end{bmatrix} = \begin{bmatrix} \cos\beta & 0 & \sin\beta & 0 \\ 0 & 1 & 0 & 0 \\ -\sin\beta & 0 & \cos\beta & 0 \\ 0 & 0 & 0 & 1 \end{bmatrix} \begin{bmatrix} x \\ y \\ z \\ 1 \end{bmatrix}. \tag{8.79}$$

- For rotation around the x axis, we have

$$\begin{bmatrix} x' \\ y' \\ z' \\ 1 \end{bmatrix} = \begin{bmatrix} 1 & 0 & 0 & 0 \\ 0 & \cos\gamma & -\sin\gamma & 0 \\ 0 & \sin\gamma & \cos\gamma & 0 \\ 0 & 0 & 0 & 1 \end{bmatrix} \begin{bmatrix} x \\ y \\ z \\ 1 \end{bmatrix}. \tag{8.80}$$

Here is an example: Find the resultant transformation matrix describing the rotation ϕ of any point **p** in the plane about the point $\mathbf{p}_c$. First, translate **p** so that $\mathbf{p}_c$ is at the origin, as described by the matrix equation

$$\mathbf{P}' = \begin{bmatrix} 1 & 0 & -x_c \\ 0 & 1 & -y_c \\ 0 & 0 & 1 \end{bmatrix} \mathbf{P}, \tag{8.81}$$

where

$$\mathbf{P} = \begin{bmatrix} x \\ y \\ 1 \end{bmatrix} \text{ and } \mathbf{P}' = \begin{bmatrix} x' \\ y' \\ 1 \end{bmatrix}. \tag{8.82}$$

Next, rotate the result through ϕ about the origin, producing

$$\mathbf{P}' = \begin{bmatrix} \cos\phi & -\sin\phi & 0 \\ \sin\phi & \cos\phi & 0 \\ 0 & 0 & 1 \end{bmatrix} \begin{bmatrix} 1 & 0 & -x_c \\ 0 & 1 & -y_c \\ 0 & 0 & 1 \end{bmatrix} \mathbf{P}. \tag{8.83}$$

Finally, translate the rotated plane back to its original position:

$$\mathbf{P}' = \begin{bmatrix} 1 & 0 & x_c \\ 0 & 1 & y_c \\ 0 & 0 & 1 \end{bmatrix} \begin{bmatrix} \cos\phi & -\sin\phi & 0 \\ \sin\phi & \cos\phi & 0 \\ 0 & 0 & 1 \end{bmatrix} \begin{bmatrix} 1 & 0 & -x_c \\ 0 & 1 & -y_c \\ 0 & 0 & 1 \end{bmatrix} \mathbf{P}. \tag{8.84}$$

If we carry out the indicated matrix multiplication, then we should find that the matrix equation representing the composite transformation, in the order given, is

$$\mathbf{P}' = \begin{bmatrix} \cos\phi & -\sin\phi & -x_c\cos\phi + y_c\sin\phi + x_c \\ \sin\phi & \cos\phi & -x_c\sin\phi - y_c\cos\phi + y_c \\ 0 & 0 & 1 \end{bmatrix} \mathbf{P}. \tag{8.85}$$

In terms of the coordinates,

$$\begin{aligned} x' &= (x - x_c)\cos\phi - (y - y_c)\sin\phi + x_c, \\ y' &= (x - x_c)\sin\phi + (y - y_c)\cos\phi + y_c. \end{aligned} \tag{8.86}$$

If $x_c = y_c = 0$, then the equations describe a simple rotation about the origin.

8.8 Kinematics

This section is an excerpt from Geometric Transformations for 3D Modeling, *2007. It is relatively brief, but it offers a view of the role of vectors in translation and rotation computations related to animation, robotics, and similar fields.*

Visualize a collection of rigid solid bodies joined so that only certain constrained motions are possible. Such an arrangement is a mechanism. Kinematics is the study of the motion of a mechanism. A serial link mechanism (this could be the manipulator arm of a robot, for example) is a sequence of links connected by actuated joints. We will examine only a simple, highly idealized serial link mechanism. However, keep in mind that mechanisms with extremely complex joint geometries and motions are possible.

A serial link mechanism with n links and n joints has n degrees of freedom. Usually we define a reference base with Link 1 connected to it through Joint 1. There is no joint at the end of the last link. The rigid links maintain a fixed geometric relationship between the joints at each end of the link. We face two problems: First, given a set of joint motions (expressed as relative transformations), find the resultant position of the end of the last link; second, given an end position, find an appropriate corresponding set of joint motions.

Here is a simple example, consisting of a planar assembly of two links and two joints constrained to joint rotations in the plane (Figure 8.18).

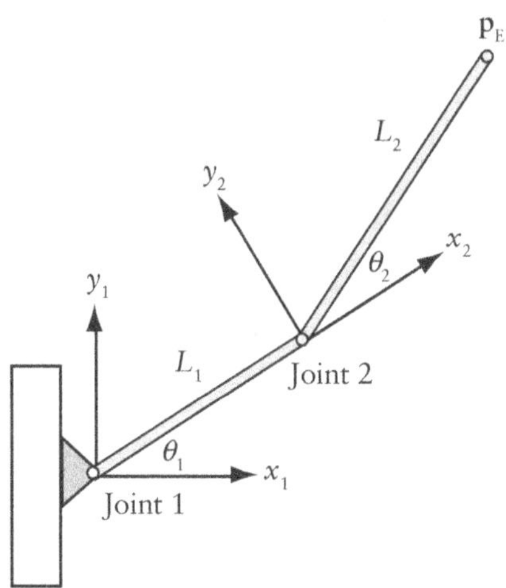

Figure 8.18 An articulated mechanism

Our first job is to express the relationships between the links. To do this we establish a coordinate frame for each link. The frame for Link 1 we fix relative to the base with its origin at Joint 1. The frame for Link 2 we fix relative to Link 1 with its origin at Joint 2 and its x axis collinear with the length of Link 1. The relationship between Link 1 and Link 2 we express as a product of homogeneous transformations, $\mathbf{H}_1$, relating the coordinate frame of Link 2 to the coordinate frame of Link 1. For our example, this is simply the product of a rotation and translation. Thus,

$$\begin{aligned}\mathbf{H}_1 &= [\text{Rotation } \theta_1][\text{Translation } L_1] \\ &= \begin{bmatrix} \cos\theta_1 & -\sin\theta_1 & 0 \\ \sin\theta_1 & \cos\theta_1 & 0 \\ 0 & 0 & 1 \end{bmatrix} \begin{bmatrix} 1 & 0 & L_1 \\ 0 & 1 & 0 \\ 0 & 0 & 1 \end{bmatrix}.\end{aligned} \tag{8.87}$$

Finally, we must relate the position of the end of Link 2, E, to the coordinate frame of Link 2. This is nothing more than the product, $\mathbf{H}_2$, of rotation and translation matrices describing motion at Joint 2:

$$\begin{aligned}\mathbf{H}_2 &= [\text{Rotation } \theta_2][\text{Translation } L_2] \\ &= \begin{bmatrix} \cos\theta_2 & -\sin\theta_2 & 0 \\ \sin\theta_2 & \cos\theta_2 & 0 \\ 0 & 0 & 1 \end{bmatrix} \begin{bmatrix} 1 & 0 & L_2 \\ 0 & 1 & 0 \\ 0 & 0 & 1 \end{bmatrix}.\end{aligned} \tag{8.88}$$

It follows that the position of E relative to Frame 1, or the base, is

$$\mathbf{p}_E = \mathbf{H}_1\mathbf{H}_2\mathbf{p}_0, \tag{8.89}$$

where we set $\mathbf{p}_0 = [\ 0\ \ 0\ \ 1\ ]^{\mathrm{T}}$. (Do you see why?)

In three dimensions, we might encounter links whose relative joint rotation axes are skew lines in space, perhaps with telescoping lengths and joints that twist. The number of potential motion variables is almost unlimited. We express the position of the end as a generalization of the preceding equation:

$$\begin{aligned}\mathbf{p}_E &= \mathbf{H}_1\mathbf{H}_2 \ldots \mathbf{H}_{n-1}\mathbf{H}_n\mathbf{p}_0 \\ &= \prod \mathbf{H}_i\mathbf{p}_0.\end{aligned} \tag{8.90}$$

Here, however, the $\mathbf{H}_i$ are certainly more complex than a rotation and translation.

Given the end position, E, there are usually many sets of joint motions possible to achieve it. In Figure 8.19, we see that there are two possibilities for the 2-link mechanism described above. As you would imagine, the number of sets of joint motions that will produce a given end position increases rather rapidly as the number and complexity of the individual joints increase.

A central problem in mechanism control is obtaining a solution for the joint coordinates given any net transformation $\mathbf{H}$. We usually know where we want to position the end of the mechanism, but we must find the appropriate joint coordinates describing the movement to achieve this position. Thus, given any transformation $\mathbf{H}$, find the Euler angles ϕ, θ, and ψ.

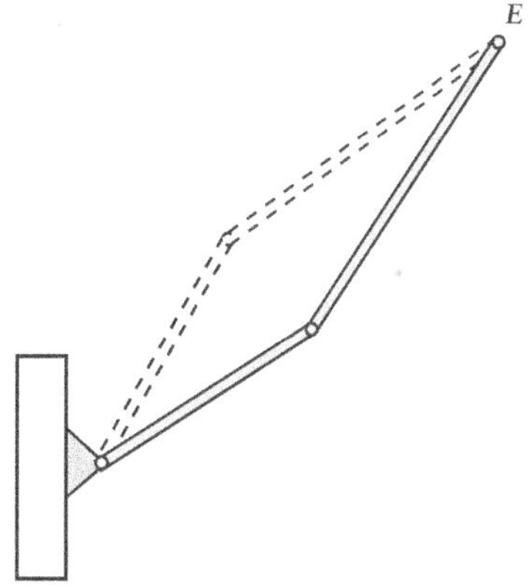

Figure 8.19 Multiple joint positions for a given end position

In other words, given the numerical values of the elements of H, what are the corresponding values of ϕ, θ, and ψ? Recall that we have already solved a similar problem: Given **R**, find ϕ, θ, and ψ. The presence of translations is only a minor complication.

8.9 Exercises

8.1. Find appropriate translation pairs Δx and Δy such that the algebraic equation of a straight line in slope-intercept form is simplified (that is, the line is translated so that it passes through the origin).

8.2. Using vectors, show that a translated line is parallel to its original position.

8.3. How far is each point moved when translated $x' = x + a$, $y' = y + b$, $z' = z + c$?

8.4. Find the equations of the translation that sends (–4, 3) into (1, 2).

8.5. Find the transformed equations of the following equations subject to the transformation $x' = x + 2$, $y' = y - 1$:

a. $x + y - 1 = 0$

b. $x^2 + y^2 = 4$

c. $y = 2x^2$

8.6. Express each of the following translations in vector form, giving the components of each vector:

a. $x' = x + 5, \quad y' = y$

b. $x' = x - 3, \quad y' = y + 2$

c. $x' = x, \quad y' = y, \quad z' = z$

d. $x' = x - 1, \quad y' = y, \quad z' = z + 6$

e. $x' = x + a, \quad y' = y + b, \quad z' = z + c$

8.7. Find the rotation matrix for rotations of (a) 30°, (b) 45°, (c) 90°, and (d) 180° in the xy plane about the origin.

8.8. Compute the determinant of each of the rotation matrices found for the exercise above.

8.9. Find the algebraic eqűations for rotating a point x, y through an angle θ about x_c, y_c.

8.10. Consider successive rotations about two different points. Is the outcome independent of the order in which we perform the rotations? Explain your answer.

8.11. Verify

$$\mathbf{R}_{\varphi\theta\phi} = \begin{bmatrix} \cos\theta\cos\phi & -\cos\theta\sin\phi & \sin\theta \\ \cos\varphi\sin\phi + \sin\varphi\sin\theta\cos\phi & \cos\varphi\cos\phi - \sin\varphi\sin\theta\sin\phi & -\sin\varphi\cos\theta \\ \sin\varphi\sin\phi - \cos\varphi\sin\theta\cos\phi & \sin\varphi\cos\phi + \cos\varphi\sin\theta\sin\phi & \cos\varphi\cos\theta \end{bmatrix}$$

by setting

a. $\alpha = \beta = 0$

b. $\alpha = \gamma = 0$

c. $\beta = \gamma = 0$

8.12. Find the equation of the following curves after the rotation transformation $x' = (x-y)/\sqrt{2}$, $y' = (x+y)/\sqrt{2}$:

a. $y = 0$

b. $y = x$

c. $y = x + 3$

d. $x = 3$

e. $x^2 + y^2 = 1$

f. $xy = 1$

g. $(x - 1)^2 + y^2 = 2$

h. $y = x^2$

i. Find the angle of rotation, ϕ.

8.13. Find the rotation matrix for very small angles of rotation. Do this for two- and three-dimensional rotations.

8.14. Show that the determinant of every translation transformation in homogeneous coordinates is equal to 1.

8.15. Show that the sum of the squares of any row or column of a proper rotation matrix is equal to 1. This is the distinguishing characteristic of a normalized matrix.

8.16. Show that the scalar product of any pair of rows (or columns) of a proper rotation matrix is zero. A matrix with this property is orthogonal.

9 More Transformations

Translation and rotation, and their combination, were discussed in depth in Chapter 8; they are briefly reviewed here. Next, this chapter introduces several more geometric transformations: scaling, shear, reflection, inversion, and projection. A discussion about how transformations are used to create sweep surfaces and solids completes this chapter. Geometric transformations rely on vectors and matrices. They are at the core of many model-building methods, such as sweep- and deformation-generated shapes, as well as methods for producing animated effects.

9.1 Translation

To review: Translation and rotation are rigid-body transformations. Rigid-body transformations preserve distance between all points of the transformed geometric object, meaning that size and shape don't change when we translate or rotate a geometric object.

The translation of a point vector **p** a distance and direction given by another vector **t** we express as

$$\mathbf{p}' = \mathbf{p} + \mathbf{t}, \tag{9.1}$$

or as matrices:

$$\mathbf{P}' = \mathbf{P} + \mathbf{T}, \tag{9.2}$$

where **P′**, **P**, and **T** are the column matrices:

$$\begin{bmatrix} p'_x \\ p'_y \\ p'_z \end{bmatrix} = \begin{bmatrix} p_x \\ p_y \\ p_z \end{bmatrix} + \begin{bmatrix} t_x \\ t_y \\ t_z \end{bmatrix}. \tag{9.3}$$

Or more usefully, in terms of homogeneous coordinates:

$$\mathbf{P}' = \mathbf{PT}, \tag{9.4}$$

where

$$\mathbf{T} = \begin{bmatrix} 1 & 0 & 0 & t_x \\ 0 & 1 & 0 & t_y \\ 0 & 0 & 1 & t_z \\ 0 & 0 & 0 & 1 \end{bmatrix}. \tag{9.5}$$

9.2 Rotation

Here we summarize rotation about one or more of the principal axes. The so-called right-hand rule defines positive rotations.

As we learned in Chapter 8, rotation about the z axis produces a result identical to a rotation about the origin in the xy plane. But we must also account for the z coordinate. This is the result:

$$\begin{aligned} x' &= x\cos\phi - y\sin\phi, \\ y' &= x\sin\phi + y\cos\phi, \\ z' &= z. \end{aligned} \tag{9.6}$$

The rotation matrix is

$$\mathbf{R}_\phi = \begin{bmatrix} \cos\phi & -\sin\phi & 0 \\ \sin\phi & \cos\phi & 0 \\ 0 & 0 & 1 \end{bmatrix}. \tag{9.7}$$

For rotation about the x or y axis, the matrices are

$$\mathbf{R}_\varphi = \begin{bmatrix} 1 & 0 & 0 \\ 0 & \cos\varphi & -\sin\varphi \\ 0 & \sin\varphi & \cos\varphi \end{bmatrix} \tag{9.8}$$

and

$$\mathbf{R}_\theta = \begin{bmatrix} \cos\theta & 0 & \sin\theta \\ 0 & 1 & 0 \\ -\sin\theta & 0 & \cos\theta \end{bmatrix}. \tag{9.9}$$

If the rotation we are interested in is the product of several rotations, we first identify the rotation sequence and then assemble the appropriate matrices in the desired transformation order to obtain the final rotation matrix.

Here is one possible sequence: rotation first about the z axis, then the y axis, and finally the x axis:

$$\mathbf{P}' = \mathbf{R}_\varphi \mathbf{R}_\theta \mathbf{R}_\phi \mathbf{P}. \tag{9.10}$$

This produces

$$\mathbf{R}_{\varphi\theta\phi} = \begin{bmatrix} \cos\theta\cos\phi & -\cos\theta\sin\phi & \sin\theta \\ \cos\varphi\sin\phi + \sin\varphi\sin\theta\cos\phi & \cos\varphi\cos\phi - \sin\varphi\sin\theta\sin\phi & -\sin\varphi\cos\theta \\ \sin\varphi\sin\phi - \cos\varphi\sin\theta\cos\phi & \sin\varphi\cos\phi + \cos\varphi\sin\theta\sin\phi & \cos\varphi\cos\theta \end{bmatrix} \tag{9.11}$$

and in the matrix of homogeneous coordinates

$$\mathbf{R}_{\varphi\theta\phi} = \begin{bmatrix} \cos\theta\cos\phi & -\cos\theta\sin\phi & \sin\theta & 0 \\ \cos\varphi\sin\phi + \sin\varphi\sin\theta\cos\phi & \cos\varphi\cos\phi - \sin\varphi\sin\theta\sin\phi & -\sin\varphi\cos\theta & 0 \\ \sin\varphi\sin\phi - \cos\varphi\sin\theta\cos\phi & \sin\varphi\cos\phi + \cos\varphi\sin\theta\sin\phi & \cos\varphi\cos\theta & 0 \\ 0 & 0 & 0 & 1 \end{bmatrix}. \tag{9.12}$$

9.3 Scaling

Scaling transformations change the size and shape of geometric objects in ways that are easy to visualize. We can apply them in two- and three-dimensional spaces, and higher spaces, too, for that matter. In addition to size and shape sculpting, they are useful in modeling the effects of physical forces on a body and in generating interesting animation effects.

A uniform scaling (or isotropic scaling) is a transformation that changes the size of a geometric object, but not its shape. We can perform this transformation about the origin or some other point. We define it by a scale factor S and a center about which it takes place:

- If $s > 1$, then the object operated on increases in size (expansion).
- If $s < 1$, then the object reduces in size (contraction).
- If $s = 1$, then the object is unchanged.

Here are some geometric properties that are invariant when a shape undergoes a uniform scaling transformation:

- Corresponding angles are unchanged.
- Parallel lines remain parallel.
- Collinear points remain collinear.
- Orientation remains unchanged.

Clearly, distance is not preserved in scaling transformations.

We can apply a uniform scaling to expand or contract the two- or three-dimensional coordinate systems in which we define geometric objects. We can choose to do this transformation about some fixed point or center (Figure 9.1). If that point is the origin, then the transformation is a homogeneous isotropic (uniform) scaling. Figure 9.1a shows an expansion s of a vector space, and Figure 9.1b shows a contraction $1/s$.

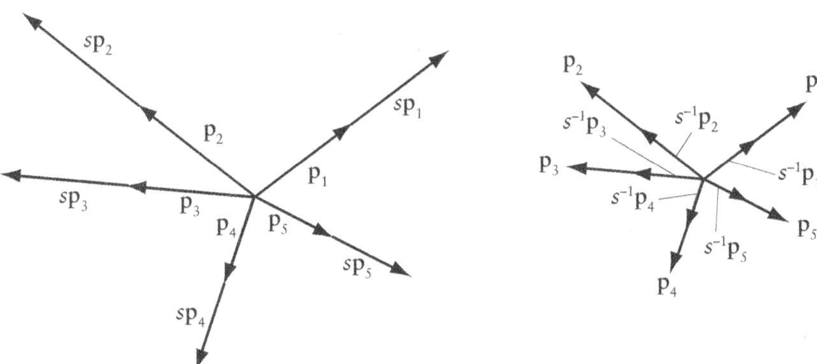

Figure 9.1 Uniform expansion and contraction

Although uniform scaling does not affect angle size, it does change distances. However, the product of a uniform scaling and a rigid-body transformation produces similar figures; it multiplies all distances or lengths by the same number, s, where s is the ratio of the transformation. Physicists and engineers often refer to this ratio as strain.

Now let's look at a uniform scaling that fixes the origin. The Cartesian equations for a scaling fixing the origin are

$$x' = sx, \quad y' = sy, \quad z' = sz. \tag{9.13}$$

In terms of vectors, this is simply a scalar multiplication: $\mathbf{p}' = s\mathbf{p}$. In matrix form we write

$$\mathbf{P}' = \mathbf{SP}, \tag{9.14}$$

where the transformation matrix is

$$\mathbf{S} = \begin{bmatrix} s & 0 & 0 \\ 0 & s & 0 \\ 0 & 0 & s \end{bmatrix}, \tag{9.15}$$

or the homogeneous matrix

$$\mathbf{S} = \begin{bmatrix} s & 0 & 0 & 0 \\ 0 & s & 0 & 0 \\ 0 & 0 & s & 0 \\ 0 & 0 & 0 & 1 \end{bmatrix}. \tag{9.16}$$

We see that the diagonal elements control the scaling transformation and that the inverse of any scaling matrix $\mathbf{S}$ is simply

$$\mathbf{S}^{-1} = \begin{bmatrix} 1/s & 0 & 0 \\ 0 & 1/s & 0 \\ 0 & 0 & 1/s \end{bmatrix}, \tag{9.17}$$

or

$$\mathbf{S}^{-1} = \begin{bmatrix} 1/s & 0 & 0 & 0 \\ 0 & 1/s & 0 & 0 \\ 0 & 0 & 1/s & 0 \\ 0 & 0 & 0 & 1 \end{bmatrix}, \tag{9.18}$$

which reverses $\mathbf{S}$, restoring the original size of an object scaled by $\mathbf{S}$. This is easy to demonstrate. Just do the matrix multiplication:

$$\mathbf{SS}^{-1} = \mathbf{I}. \tag{9.19}$$

Let's take a closer look at just what we mean by uniform expansion. We can test the following hypothesis: If every pair of points separated by a distance d before a uniform scaling with $s > 1$ is separated by sd after the scaling, then the space has undergone a uniform expansion, and similarly for a uniform contraction with $s < 1$. We'll use vector algebra to prove it (Figure 9.2).

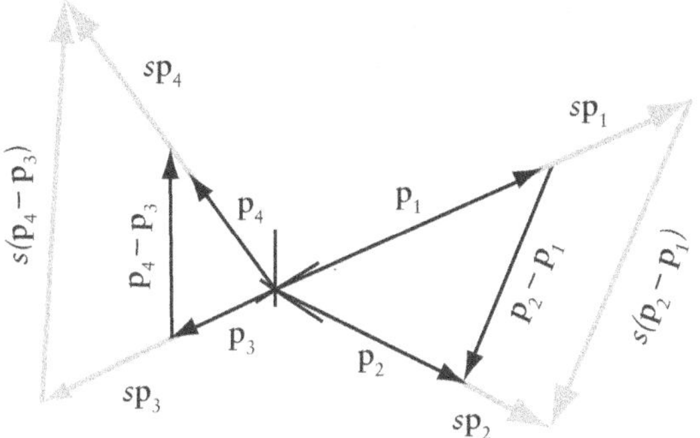

Figure 9.2 Uniform expansion

The distance between two points $\mathbf{p}_1$ and $\mathbf{p}_2$ is $|\mathbf{p}_2 - \mathbf{p}_1|$, where $d = |\mathbf{p}_2 - \mathbf{p}_1|$. After an expansion s, the distance between $\mathbf{p}_1$ and $\mathbf{p}_2$ is sd, since $|\mathbf{p}'_2 - \mathbf{p}'_1| = s|\mathbf{p}_2 - \mathbf{p}_1|$. Apply a similar argument to any pair of points where $|\mathbf{p}_i - \mathbf{p}_j| = d$.

Another form of scaling is the centerless uniform expansion. We refer to a fixed point or center of expansion under a scaling transformation, but this point is merely an artifact of the Cartesian formulation of this transformation. Here is a different way of looking at uniform expansion.

Using coordinate-free vector geometry, we can demonstrate that the expansion under a uniform scaling s looks the same from any point $\mathbf{p}$ independent of the fixed point or center of expansion used in the Cartesian formulation of the transformation. An observer at $\mathbf{p}$ stakes out, or in some way marks, a point at a distance and direction $\mathbf{q}$ from $\mathbf{p}$, and another in the opposite direction, $-\mathbf{q}$ (Figure 9.3). The observer is thus at the midpoint of a line joining points $\mathbf{p} + \mathbf{q}$ and $\mathbf{p} - \mathbf{q}$ whose length is $L = 2|\mathbf{q}|$.

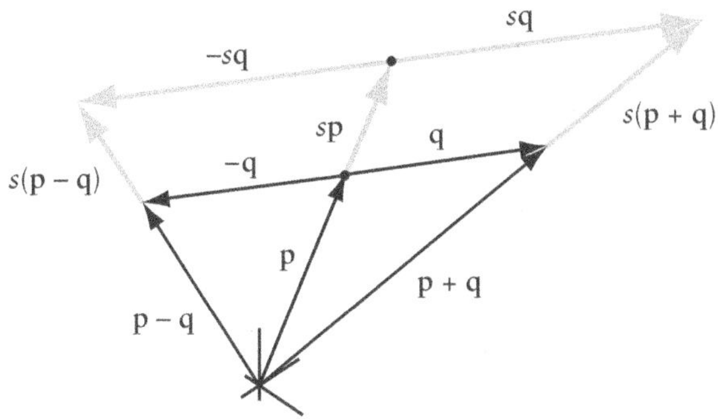

Figure 9.3 Centerless uniform expansion

After a uniform expansion, s, the observer is at $\mathbf{p}' = s\mathbf{p}$, and the two staked points are now at $\mathbf{p}' + \mathbf{q}' = s(\mathbf{p} + \mathbf{q})$ and $\mathbf{p}' - \mathbf{q}' = s(\mathbf{p} - \mathbf{q})$. The length of the line joining them is $L' = 2s|\mathbf{q}|$. The observer compares L with L' and finds $L'/L = s$. Clearly, this is true no matter what the observer's position (including at the so-called center of expansion) and no matter where the observer places the stakes. Measurements made at or relative to any point **p** before and after expansion will yield the same uniform scale factor s. A good metaphor to represent a centerless expansion is the points on the surface of a balloon at successive stages of inflation. All move uniformly farther away from each other, and there is no privileged center point.

Let's now look at nonuniform (also known as anisotropic) scaling in the xy plane to understand how the expansion ratio varies with direction. Assume that the dilation axes correspond to the x and y axes. Let s_ϕ denote the ratio in the direction ϕ with respect to the x axis (Figure 9.4).

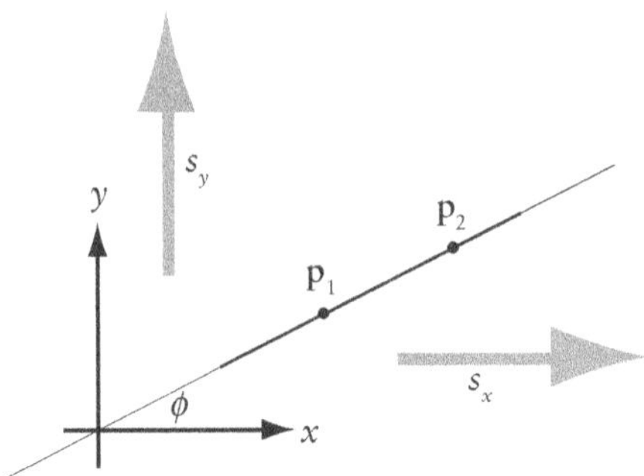

Figure 9.4 Nonuniform scaling in the plane

The strategy here is to compare distances between a pair of points, say $\mathbf{p}_1$ and $\mathbf{p}_2$, on a line through the origin, before and after a nonuniform scaling, using $s_\phi = L'/L$. The distance between two points $\mathbf{p}_1$ and $\mathbf{p}_2$ is

$$L = \sqrt{(x_2 - x_1)^2 + (y_2 - y_1)^2}. \tag{9. 20}$$

After a nonuniform scaling with ratios s_x and s_y, the distance between these same points is

$$L' = \sqrt{(x_2' - x_1')^2 + (y_2' - y_1')^2}. \tag{9.21}$$

Substituting from Equation 9.20 for x' and y', we find

$$L' = \sqrt{s_x^2(x_2 - x_1)^2 + s_y^2(y_2 - y_1)^2}, \tag{9.22}$$

so that

$$\begin{aligned} s_\phi &= \frac{L'}{L} \\ &= \sqrt{\frac{s_x^2(x_2 - x_1)^2 + s_y^2(y_2 - y_1)^2}{(x_2 - x_1)^2 + (y_2 - y_1)^2}}. \end{aligned} \tag{9.23}$$

This equation hides the effect of orientation. We can reveal it by using the relationship $y = x\tan\phi$. Appropriately substituting this into Equation 9.23, we obtain

$$s_\phi = \sqrt{s_x^2 \cos^2\phi + s_y^2 \sin^2\phi}. \tag{9.24}$$

This is much better. Since s_x and s_y are orthogonal, we immediately see by inspection the maximum and minimum scaling and the sought-after relationship to direction, all in a single expression. We also recognize that a graph of s_ϕ and ϕ produces an ellipse (Figure 9.5), where s_x and s_y are the major and minor axes depending on their relative magnitudes.

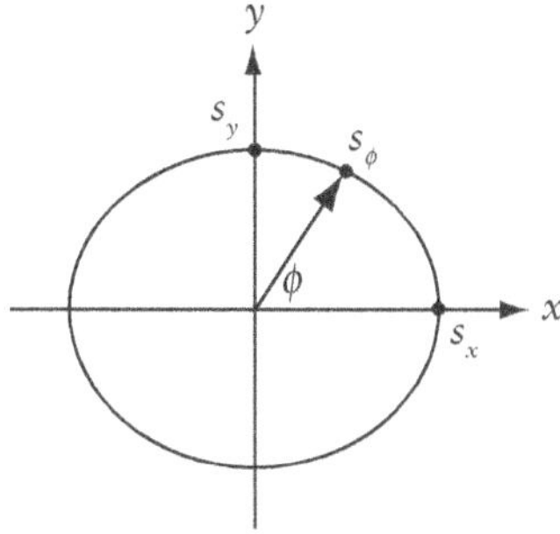

Figure 9.5 Variation of k_ϕ in the plane

Now let's look at some slightly more complex scaling. Although we will continue to focus our attention on these transformations as they apply to geometry in the plane, the extension to three dimensions is fairly obvious and direct.

There is no reason to constrain the scaling axes to a collinear correspondence with the coordinate axes. We can consider just as easily two mutually orthogonal scalings s_1 and s_2, where s_1 makes an angle α with the x axis (Figure 9.6). In this case, the final transformation matrix **H** is a product of three other matrices: $\mathbf{R}_\alpha^{-1}$ rotates s_1 into the x axis, **S** is the scaling, and $\mathbf{R}_\alpha$ rotates s_1 back to its original orientation. Thus,

$$\mathbf{H} = \mathbf{R}_\alpha \mathbf{S} \mathbf{R}_\alpha^{-1}, \tag{9.25}$$

or

$$\mathbf{H} = \begin{bmatrix} s_1\cos^2\alpha + s_2\sin^2\alpha & (s_1 - s_2)\sin\alpha\cos\alpha & 0 \\ (s_1 - s_2)\sin\alpha\cos\alpha & s_1\sin^2\alpha + s_2\cos^2\alpha & 0 \\ 0 & 0 & 1 \end{bmatrix}. \tag{9.26}$$

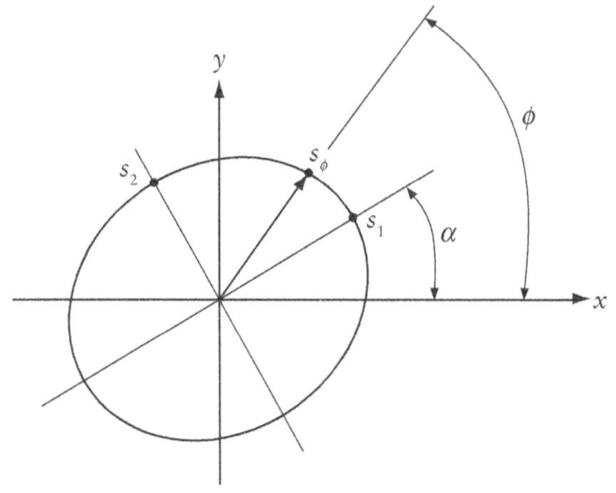

Figure 9.6 General orthogonal scaling in the plane

The net scaling s_ϕ in a direction ϕ lies on an ellipse rotated by α with respect to the coordinate axes, where s_1 and s_2 are the major and minor axes of the ellipse, again depending on their relative magnitudes.

We can also remove another constraint and generalize this transformation even further. If we no longer require an orthogonal relationship between s_1 and s_2, then we must specify the orientation of each one. Let α_1 and α_2 denote the orientation of s_1 and s_2, respectively, relative to the x axis (Figure 9.7). Then we write

$$\mathbf{H} = \mathbf{R}_{\alpha 2}\mathbf{S}_2\mathbf{R}_{\alpha 2}^{-1}\mathbf{R}_{\alpha 1}\mathbf{S}_1\mathbf{R}_{\alpha 1}^{-1}, \tag{9.27}$$

where

$$\mathbf{S}_1 = \begin{bmatrix} s_1 & 0 & 0 \\ 0 & 1 & 0 \\ 0 & 0 & 1 \end{bmatrix} \text{ and } \mathbf{S}_2 = \begin{bmatrix} 1 & 0 & 0 \\ 0 & s_2 & 0 \\ 0 & 0 & 1 \end{bmatrix}. \tag{9.28}$$

Now determining the effect of orientation becomes far more complex, as is the case for finding maximum and minimum dilations. We will not attempt this here. It is, however, a challenge worth the effort.

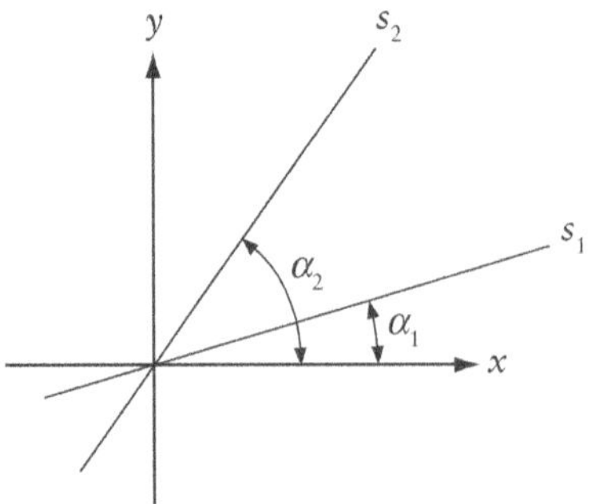

Figure 9.7 General nonorthogonal scaling in the plane

Now we move on to nonuniform scalings fixing an arbitrary point, say $\mathbf{p}_c$. To do this we simply introduce a translation $\mathbf{T}_c$ and its reverse $\mathbf{T}_{-c}$. The general transformation matrix then becomes

$$\mathbf{H} = \mathbf{T}_c\mathbf{S}\mathbf{T}_{-c}, \tag{9.29}$$

or

$$\mathbf{H} = \begin{bmatrix} 1 & 0 & 0 & x_c \\ 0 & 1 & 0 & y_c \\ 0 & 0 & 1 & z_c \\ 0 & 0 & 0 & 1 \end{bmatrix} \begin{bmatrix} s_x & 0 & 0 & 0 \\ 0 & s_y & 0 & 0 \\ 0 & 0 & s_z & 0 \\ 0 & 0 & 0 & 1 \end{bmatrix} \begin{bmatrix} 1 & 0 & 0 & -x_c \\ 0 & 1 & 0 & -y_c \\ 0 & 0 & 1 & -z_c \\ 0 & 0 & 0 & 1 \end{bmatrix}, \tag{9.30}$$

so that

$$\mathbf{H} = \begin{bmatrix} s_x & 0 & 0 & (1-s_x)x_c \\ 0 & s_y & 0 & (1-s_y)y_c \\ 0 & 0 & s_z & (1-s_z)z_c \\ 0 & 0 & 0 & 1 \end{bmatrix}. \tag{9.31}$$

9.4 Shear

A slippery deck of cards serves as a physical model of the shear transformation. Place a deck of cards on a table. Push down on the deck with sufficient pressure to keep it intact; then push horizontally to produce the deformation shown in Figure 9.8. If we assume that the bottom card is fixed, then each card moves horizontally a distance directly proportional to its vertical position in the deck. The top card moves the farthest. This is analogous to the shear transformation of sets of points in a plane or three-dimensional space.

Figure 9.8 A slippery deck of cards

Here are the equations for a shear transformation in the xy plane that fixes the x axis (Figure 9.9a):

$$\begin{aligned} x' &= x + v_y y, \\ y' &= y, \\ z' &= z, \end{aligned} \tag{9.32}$$

where v is the shear coefficient.

This transformation in matrix form is

$$\mathbf{P}' = \mathbf{V}_y \mathbf{P}, \tag{9.33}$$

where

$$\mathbf{V}_y = \begin{bmatrix} 1 & v_y & 0 \\ 0 & 1 & 0 \\ 0 & 0 & 1 \end{bmatrix}, \quad \mathbf{P} = \begin{bmatrix} x \\ y \\ z \end{bmatrix}, \tag{9.34}$$

or with homogeneous coordinates

$$\mathbf{V}_y = \begin{bmatrix} 1 & v_y & 0 & 0 \\ 0 & 1 & 0 & 0 \\ 0 & 0 & 1 & 0 \\ 0 & 0 & 0 & 1 \end{bmatrix}, \qquad \mathbf{P} = \begin{bmatrix} x \\ y \\ z \\ 1 \end{bmatrix}. \tag{9.35}$$

We can express the shear coefficient v_y in terms of the angle ϕ (Figure 9.9b). Thus, $v_y = -\tan\phi$. Line segments parallel to the x axis do not change length under this shear, while those parallel to the y axis change length according to the ratio $L'/L = \sqrt{1+v_y^2}$.

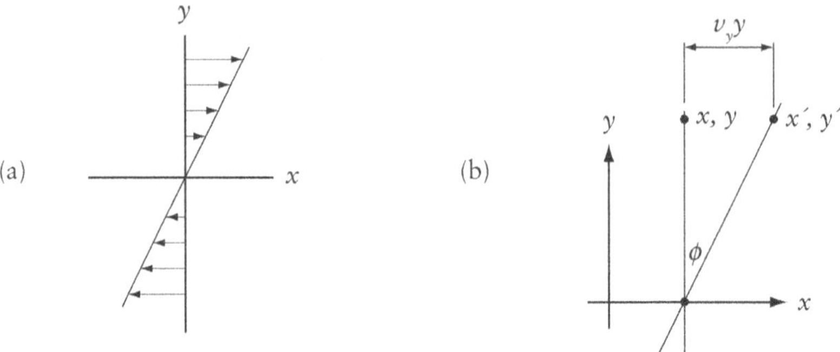

Figure 9.9 Shear in the plane fixing the x axis

Consider two kinds of shear in space that fix a plane. The formulation is simple if we choose to fix the xy plane. Study Figure 9.10, where the z axis is normal to the xy plane. Here is that slippery deck of cards again, and we are looking down at the top of it. The shear deformations we induce in Figure 9.10a and 9.10b are quite similar, but in 9.10c something different is happening.

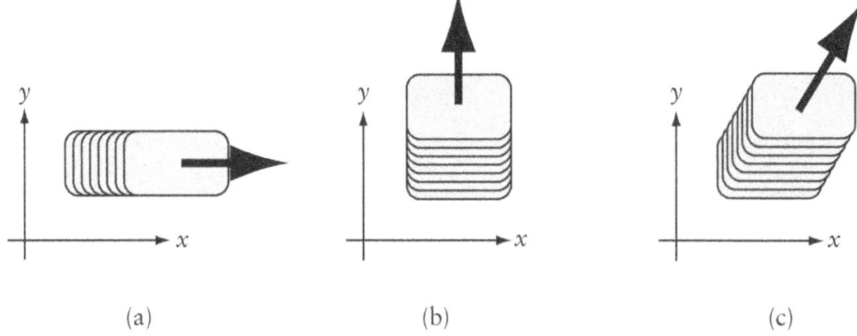

Figure 9.10 Shear in space

The two shears of 9.10a and 9.10b are now working together to produce the deformation. The transformation equations describing each of them are simply:

In Figure 9.10a,

$$\begin{aligned} x' &= x + v_1 z, \\ y' &= y, \\ z' &= z, \end{aligned} \tag{9.36}$$

and in matrix form,

$$P' = V_1P, \qquad V_1 = \begin{bmatrix} 1 & 0 & v_1 \\ 0 & 1 & 0 \\ 0 & 0 & 1 \end{bmatrix}. \tag{9.37}$$

In Figure 9.10b,

$$\begin{aligned} x' &= x, \\ y' &= y + v_2 z, \\ z' &= z, \end{aligned} \tag{9.38}$$

$$P' = V_2P, \qquad V_2 = \begin{bmatrix} 1 & 0 & 0 \\ 0 & 1 & v_2 \\ 0 & 0 & 1 \end{bmatrix}. \tag{9.39}$$

In Figure 9.10c,

$$\begin{aligned} x' &= x + v_1 z, \\ y' &= y + v_2 z, \\ z' &= z, \end{aligned} \tag{9.40}$$

$$P' = V_{12}P, \qquad V_{12} = \begin{bmatrix} 1 & 0 & v_1 \\ 0 & 1 & v_2 \\ 0 & 0 & 1 \end{bmatrix}. \tag{9.41}$$

9.5 Reflection

Reflection transformations of the kind we discuss in this section are also about symmetry. Among other uses, these transformations can facilitate the generation of symmetrical shapes and patterns. But before exploring the application of vectors and matrices, we'll begin with a general description of a reflection transformation in a plane, which we can easily extend to three dimensions.

Given any point A and a line m not through A, a reflection of A across m produces its transformed or image point A'. The line segment AA' is perpendicular to m and bisected by it (Figure 9.11). We easily demonstrate the following characteristics of a reflection:

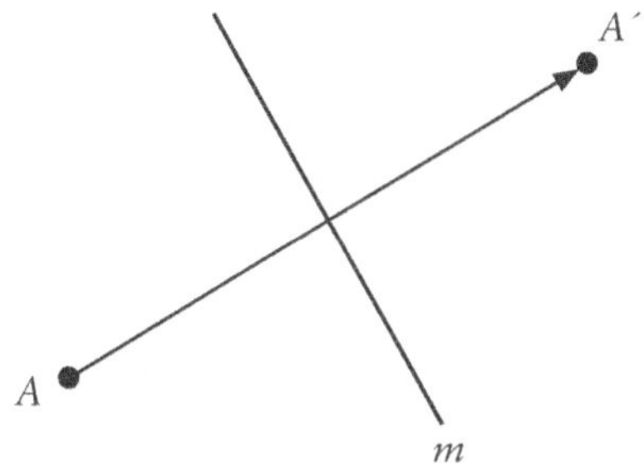

Figure 9.11 Reflection of A in m produces point A'

1. A reflection in a line in a plane reverses the orientation of a shape (Figure 9.12).
2. If a line is parallel to the axis of reflection, then so is its image. Similarly, if a line is perpendicular to the axis of reflection, then so is its image, and the line and image are collinear.
3. A reflection transformation preserves distance between points, the angle between lines, parallel lines and planes, perpendicularity, betweenness, and midpoints.

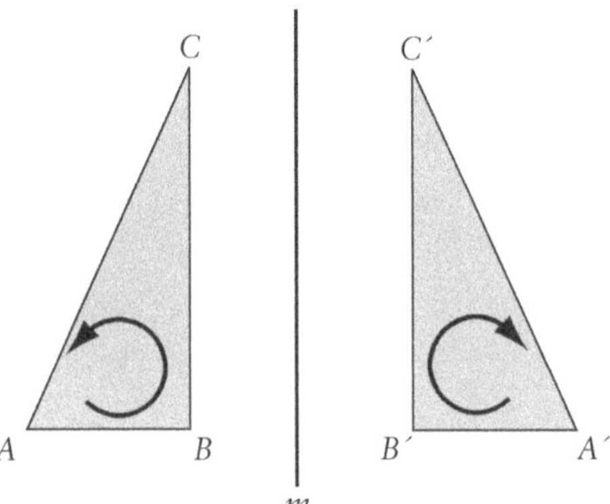

Figure 9.12 Reflection in a line in a plane reverses orientation

To find the transformation equations describing reflection about two intersecting lines, we proceed as follows: Given a figure ABC and two intersecting lines m and n in the plane, the successive reflections of ABC about m and then n produce an image $A''B''C''$, which turns out to be equivalent to a rotation about the point of intersection of the lines and through an angle equal to twice the angle between them (Figure 9.13).

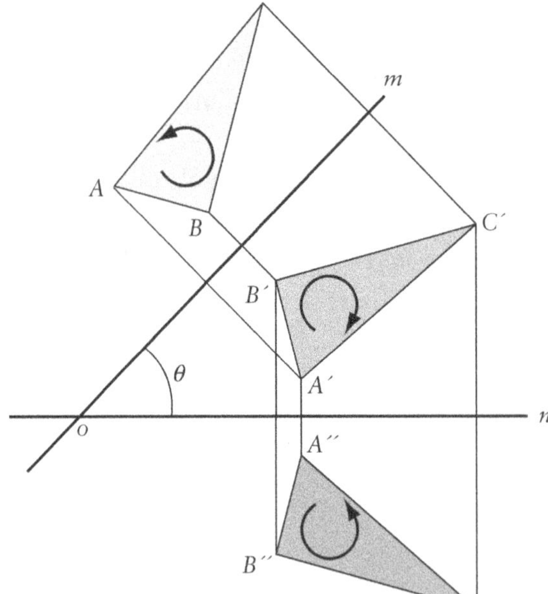

Figure 9.13 Successive reflections of *ABC* about *m* and then *n*

Reflection in the x or y axis reverses orientation (Figure 9.14). Reflection in the x axis is given by

$$\begin{aligned} x' &= x, \\ y' &= -y, \end{aligned} \tag{9.42}$$

$$\mathbf{P}' = \mathbf{R}_{fx}\mathbf{P}, \tag{9.43}$$

and in matrix form as

$$\mathbf{R}_{fx} = \begin{bmatrix} 1 & 0 \\ 0 & -1 \end{bmatrix}, \tag{9.44}$$

where $\mathbf{R}_{fx}$ denotes the reflection matrix for a reflection in the x-axis.

Reflection in the y axis is given by

$$\begin{aligned} x' &= -x, \\ y' &= y, \end{aligned} \tag{9.45}$$

and in matrix form as

$$\mathbf{P}' = \mathbf{R}_{fy}\mathbf{P}, \tag{9.46}$$

where

$$\mathbf{R}_{fy} = \begin{bmatrix} -1 & 0 \\ 0 & 1 \end{bmatrix}. \tag{9.47}$$

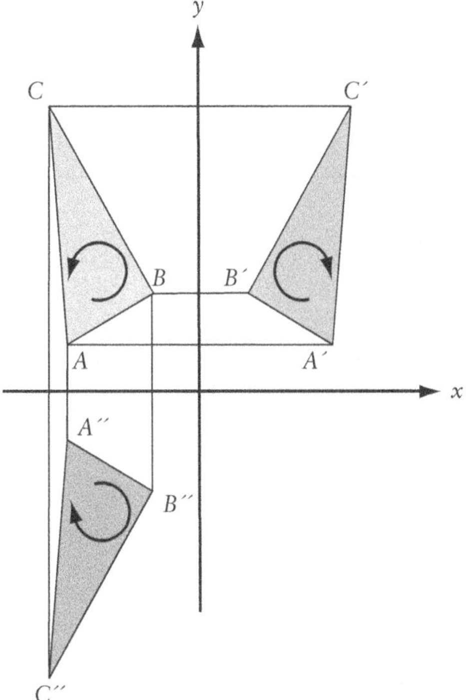

Figure 9.14 Reflection in the x or y axis

Reflection in an arbitrary line through the origin reverses orientation (Figure 9.15) and is given by

$$\begin{aligned} x' &= x\cos 2\alpha + y\sin 2\alpha, \\ y' &= x\sin 2\alpha - y\cos 2\alpha, \end{aligned} \tag{9.48}$$

$$\mathbf{P}' = \mathbf{R}_{f\alpha}\mathbf{P}, \tag{9.49}$$

$$\mathbf{R}_{f\alpha} = \begin{bmatrix} \cos 2\alpha & \sin 2\alpha \\ \sin 2\alpha & -\cos 2\alpha \end{bmatrix}. \tag{9.50}$$

This transformation consists of a rotation through $-\alpha$ that brings the reflection axis m into coincidence with the x axis, followed by a reflection across the x axis, and completed by a rotation that returns line m to its original position. It is straightforward to work through the three matrices that describe this transformation and that produce the net or final transformation matrix $\mathbf{R}_{f\alpha}$.

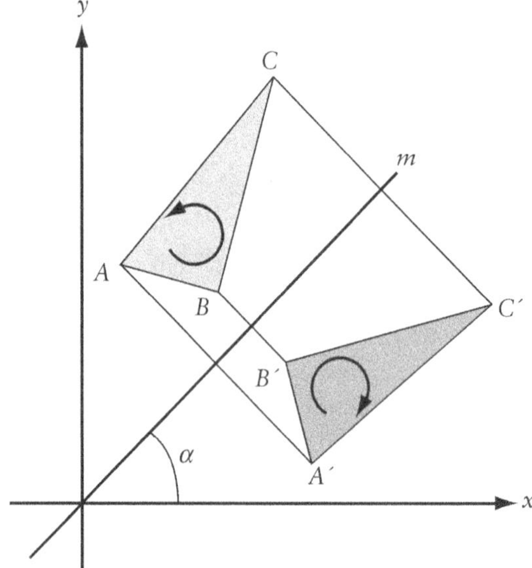

Figure 9.15 Reflection in an arbitrary line through the origin

Reflection in the xy plane ($z = 0$ plane) reverses orientation (Figure 9.16). The following set of equations defines this transformation:

$$\begin{aligned} x' &= x, \\ y' &= y, \\ z' &= -z, \end{aligned} \tag{9.51}$$

$$\mathbf{P}' = \mathbf{R}_{fz}\mathbf{P}, \tag{9.52}$$

$$\mathbf{R}_{fz} = \begin{bmatrix} 1 & 0 & 0 \\ 0 & 1 & 0 \\ 0 & 0 & -1 \end{bmatrix}. \tag{9.53}$$

It is similar for the $x = 0$ and $y = 0$ planes. To more clearly see the orientation reversal in the faces of $A'B'C'D'$, imagine looking at face $A'B'C'$ from some point outside the solid.

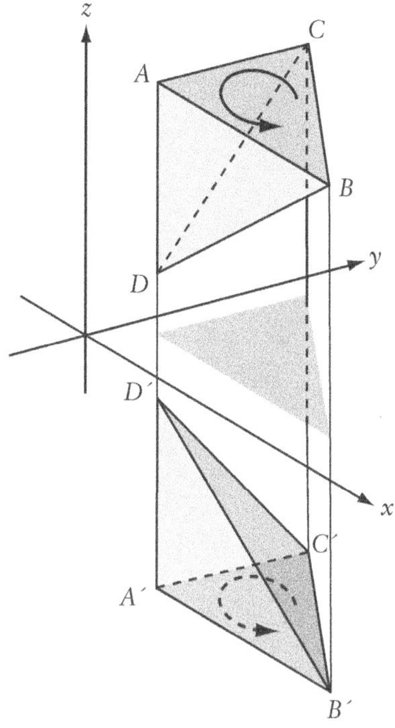

Figure 9.16 Reflection in the *xy* plane reverses orientation

9.6 Inversion

Inversion is another kind of reflection. To construct the inversion of a figure ABC through a point P, all lying in the plane, we prolong line AP so that $AP = PA'$, and similarly to obtain $BP = PB'$ (Figure 9.17). $A'B'C'$ is the inversion of ABC. By comparing the vertex sequence of these two triangles, we see that this transformation preserves orientation.

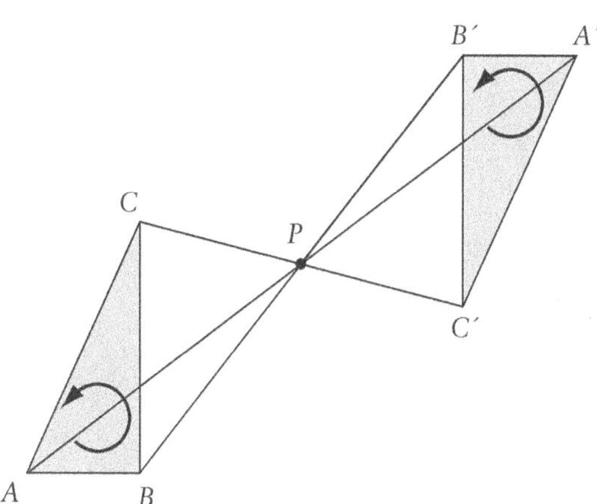

Figure 9.17 Inversion of a figure *ABC* through a point *P*

$A'B'C'$ is identical to the image produced if we rotate ABC about P. For this reason, mathematicians often call an inversion a "half-turn." Two successive half-turns about the same point are an identity transformation because they bring the figure back to its initial position. Two successive half-turns about two different points, **P** and **Q**, produce a translation (Figure 9.18).

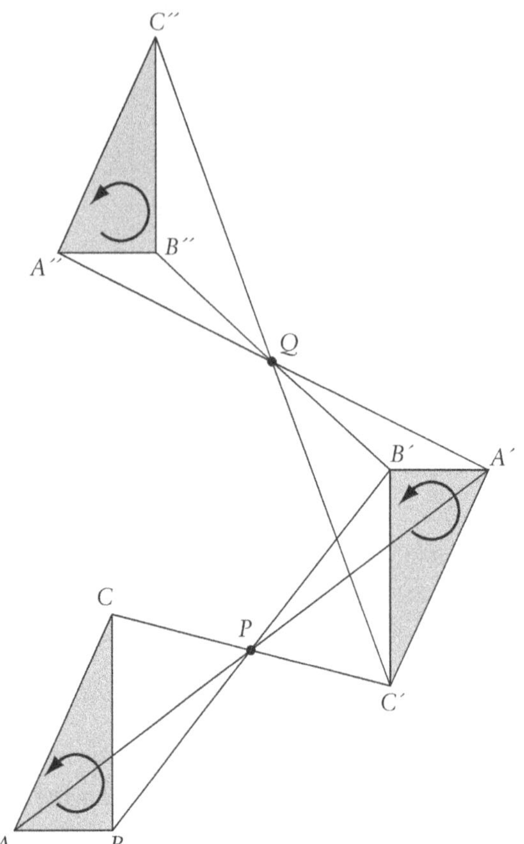

Figure 9.18 Two successive half-turns about two different points produce a translation

Inversion in the origin (Figure 9.19) is given by

$$\begin{aligned} x' &= -x, \\ y' &= -y, \end{aligned} \tag{9.54}$$

$$\mathbf{P}' = \mathbf{R}_{fo}\mathbf{P}, \tag{9.55}$$

$$\mathbf{R}_{fo} = \begin{bmatrix} -1 & 0 \\ 0 & -1 \end{bmatrix}, \tag{9.56}$$

and similarly for inversion in an arbitrary point (a, b):

$$\begin{aligned} x' &= -x + 2a, \\ y' &= -y + 2b, \end{aligned} \tag{9.57}$$

$$\mathbf{P}' = \mathbf{R}_{fab}\mathbf{P}. \tag{9.58}$$

Notice that two translations are required here. So in setting up the transformation matrix, we will use homogeneous coordinates, which allow us to combine the transformations into a single composite matrix:

$$\mathbf{R}_{fab} = \begin{bmatrix} -1 & 0 & 2a \\ 0 & -1 & 2b \\ 0 & 0 & 1 \end{bmatrix}. \tag{9.59}$$

To make this matrix multiplication work ($\mathbf{P}' = \mathbf{R}_{fab}\mathbf{P}$), we must use homogeneous coordinates for the **P** and **P′** matrices, adding an "extra" coordinate:

$$\mathbf{P} = \begin{bmatrix} x \\ y \\ 1 \end{bmatrix}, \quad \mathbf{P}' = \begin{bmatrix} x' \\ y' \\ 1 \end{bmatrix}, \tag{9.60}$$

which makes them conformable to $\mathbf{R}_{fab}$.

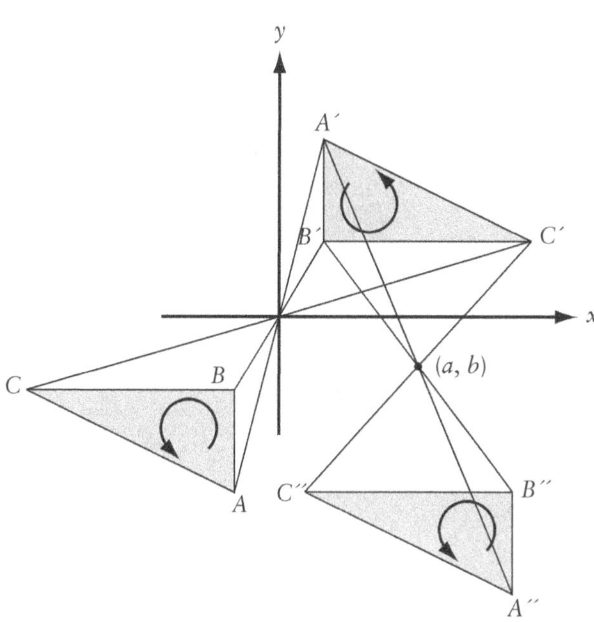

Figure 9.19 Inversion in the origin followed by inversion in the point (*a*, *b*)

Inversion in the origin or in an arbitrary point a, b, c in space reverses orientation of surfaces bounding a solid. This is shown in Figure 9.20. The tetrahedral solid *ABCD* is inverted through the origin. Compare face *ABD* with *A′B′C′*. This inversion is expressed mathematically as

$$\begin{aligned} x' &= -x, \\ y' &= -y, \\ z' &= -z, \end{aligned} \tag{9.61}$$

$$\mathbf{P}' = \mathbf{R}_{fo}\mathbf{P}, \tag{9.62}$$

$$\mathbf{R}_{fo} = \begin{bmatrix} -1 & 0 & 0 \\ 0 & -1 & 0 \\ 0 & 0 & -1 \end{bmatrix}, \tag{9.63}$$

and

$$\begin{aligned} x' &= -x + 2a, \\ y' &= -y + 2b, \\ z' &= -z + 2c, \end{aligned} \tag{9.64}$$

$$\mathbf{P}' = \mathbf{R}_{fabc}\mathbf{P}, \tag{9.65}$$

$$\mathbf{R}_{fabc} = \begin{bmatrix} -1 & 0 & 0 & 2a \\ 0 & -1 & 0 & 2b \\ 0 & 0 & -1 & 2c \\ 0 & 0 & 0 & 1 \end{bmatrix}, \quad \mathbf{P} = \begin{bmatrix} x \\ y \\ z \\ 1 \end{bmatrix}, \quad \mathbf{P}' = \begin{bmatrix} x' \\ y' \\ z' \\ 1 \end{bmatrix}. \tag{9.66}$$

Note that inversion in spaces of even dimension preserves orientation, while inversion in spaces of odd dimension reverses orientation (Figure 9.20).

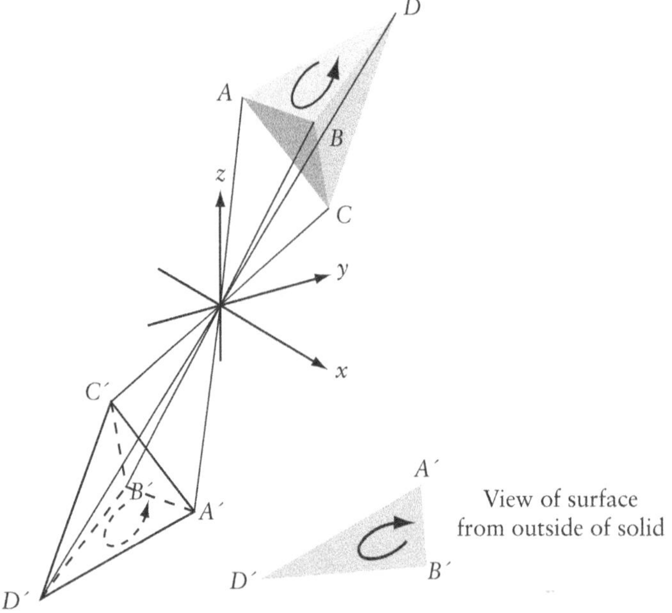

Figure 9.20 Inversion of a solid in a point reverses orientation of the bounding surfaces

9.7 Projection

Projection transformations are important for many reasons, not the least of which is that they allow us to produce and understand two-dimensional images of three-dimensional objects, and the shadows cast by these objects, which are really just projections. Figure 9.21 shows the point projection of a shape in plane A onto plane B. This section discusses several kinds of projection, including this. First, some general characteristics of projection transformations are in order.

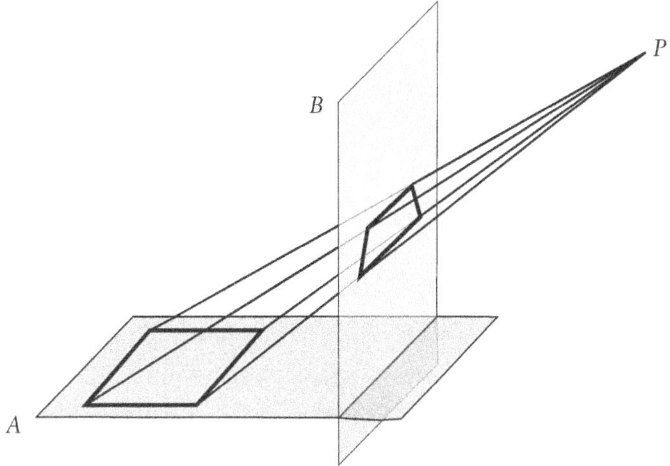

Figure 9.21 Projection of *A* onto *B* through *P*

Parallel projection uses a family of parallel lines to produce an image. In Figure 9.22 parallel lines project points in space onto the plane Π.

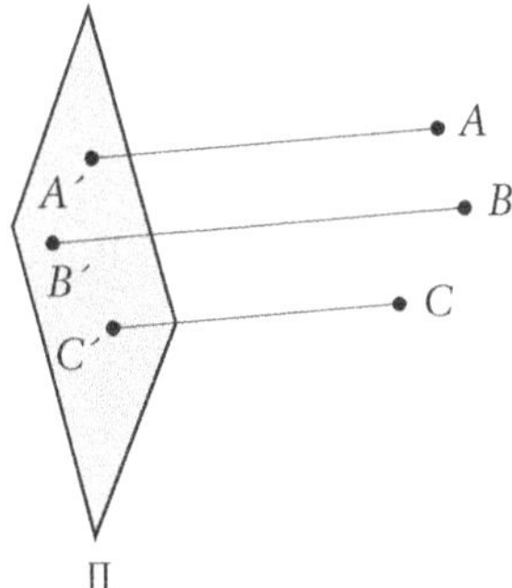

Figure 9.22 Parallel projection of points in space onto a plane

There are two kinds of parallel projection, based on the relationship between the direction of the lines of projection and the normal to the projection plane. In orthographic parallel projections, the direction of projection is normal to the projection plane, while in oblique parallel projection, it is not.

The orthogonal projection of any point **p** onto an arbitrary plane Π, given by $Ax+By+Cz+D=0$, is shown in Figure 9.23.

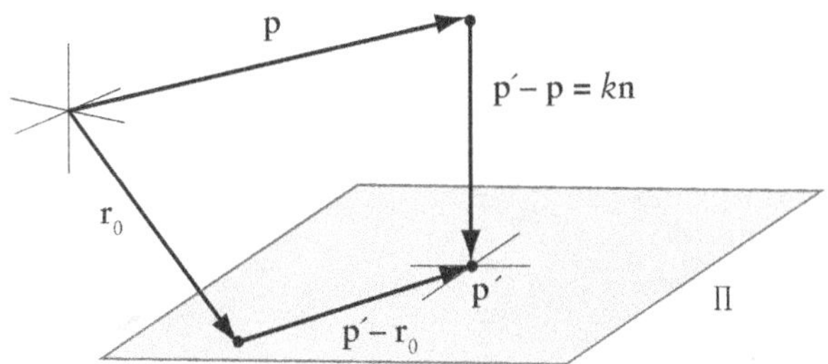

Figure 9.23 Orthogonal projection of a point onto a plane

We find the projected image $\mathbf{p}'$ by constructing a line through $\mathbf{p}$ normal to the plane. We use vector geometry to solve this problem. Find any point $\mathbf{r}_0$ on Π and the unit normal $\mathbf{n}$ using the plane's coefficients A, B, C, and D. Then

$$\mathbf{p}' = \mathbf{p} + k\mathbf{n} \tag{9.67}$$

and

$$(\mathbf{p}' - \mathbf{r}_0) \bullet (\mathbf{p}' - \mathbf{p}) = 0, \tag{9.68}$$

producing four equations in four unknowns, namely k, p'_x, p'_y, p'_z. Solving these equations and substituting appropriately yields $\mathbf{p}'$. There are several variations of this, all equally efficient.

Next, consider the parallel projection of a point onto an arbitrary plane at an oblique angle to the lines of projection (Figure 9.24). Let the unit vector $\hat{\mathbf{d}}$ denote the oblique direction of projection, $\hat{\mathbf{n}}$ the unit normal to the plane, and $\mathbf{r}_0$ any arbitrary point on the plane. *Note:* $\hat{\mathbf{n}}$ and $\mathbf{r}_0$ are functions of the plane's coefficients (or, of course, may be directly defined by vectors). We have

$$\mathbf{p}' = \mathbf{p} + k\hat{\mathbf{d}} \tag{9.69}$$

and

$$\hat{\mathbf{n}} \bullet (\mathbf{p}' - \mathbf{r}_0) = 0, \tag{9.70}$$

again producing four equations in four unknowns, namely k, p'_x, p'_y, p'_z, from which we obtain $\mathbf{p}'$.

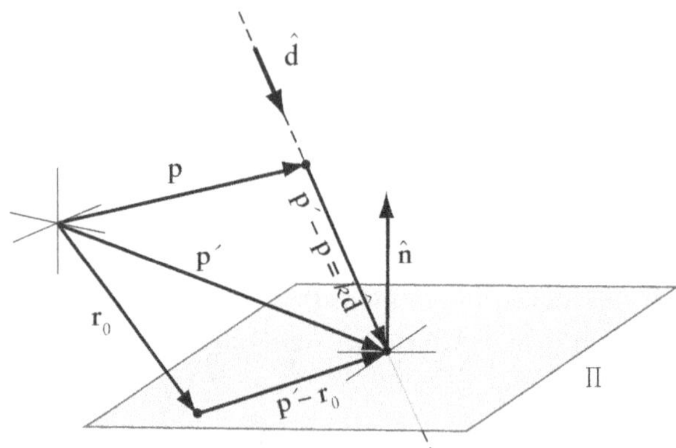

Figure 9.24 Oblique projection of a point onto a plane

Another approach is first to transform the original point set and coordinate system so that the xy plane becomes coincident with the plane of projection and then merely to execute an orthogonal projection onto it using the following transformation equation (Figure 9.25):

$$\mathbf{P}' = \begin{bmatrix} 1 & 0 & 0 & 0 \\ 0 & 1 & 0 & 0 \\ 0 & 0 & 0 & 0 \\ 0 & 0 & 0 & 1 \end{bmatrix} \mathbf{P}. \tag{9.71}$$

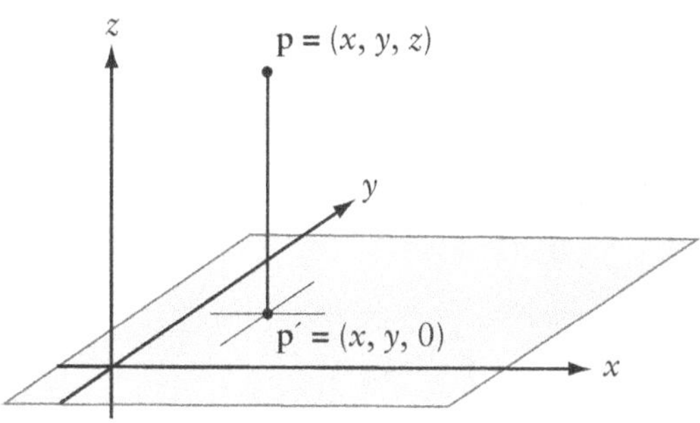

Figure 9.25 Plane of projection coincident with the *xy* plane

We compute the coordinates $\mathbf{p}'_i$ of any point $\mathbf{p}_i$ in the *xy* plane under an orthographic projection (Figure 9.26) as follows: Compute a_i, b_i, and c_i from

$$\mathbf{p}_i = \mathbf{p}_0 + a_i\mathbf{u}_1 + b_i\mathbf{u}_2 + c_i\mathbf{u}_3. \tag{9.72}$$

This equation yields three simultaneous linear equations in three unknowns a_i, b_i, and c_i. Then, compute $\mathbf{p}'_i$ from

$$\mathbf{p}'_i = \mathbf{p}_0 + a_i\mathbf{u}_1 + b_i\mathbf{u}_2. \tag{9.73}$$

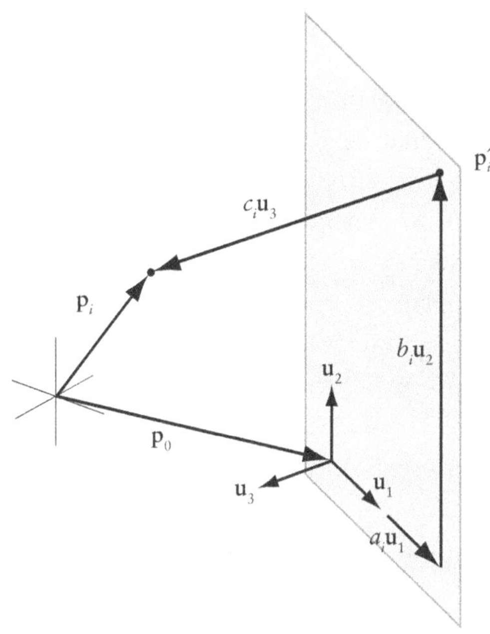

Figure 9.26 Orthographic projection

For an oblique projection compute the transformed coordinates $\mathbf{p}'_i$ of any point $\mathbf{p}_i$ (Figure 9.27) as follows: First, compute $\mathbf{a}_i$, $\mathbf{b}_i$, and $\mathbf{d}_i$ from

$$\mathbf{p}_i + d_i\mathbf{u}_4 = \mathbf{p}_0 + a_i\mathbf{u}_1 + b_i\mathbf{u}_2, \tag{9.74}$$

where $\mathbf{u}_4$ is a unit vector in the direction of the oblique projectors. This equation yields three simultaneous linear equations in three unknowns a_i, b_i, and d_i. Next, compute $\mathbf{p}'_i$ from

$$\mathbf{p}'_i = \mathbf{p}_0 + a_i\mathbf{u}_1 + b_i\mathbf{u}_2, \tag{9.75}$$

or

$$\mathbf{p}'_i = \mathbf{p}_i + d_i\mathbf{u}_4. \tag{9.76}$$

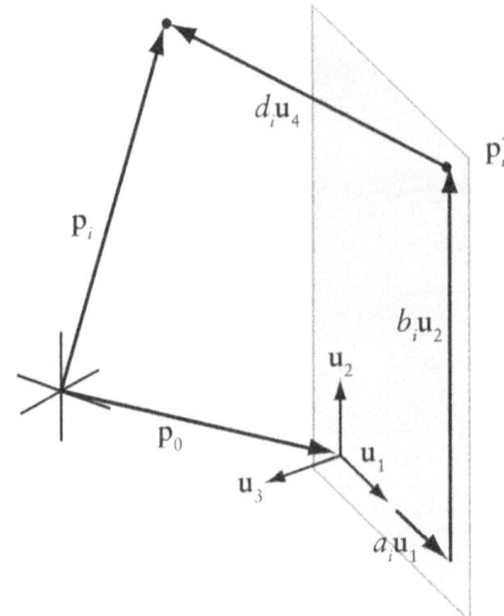

Figure 9.27 Oblique projection

Central, or perspective, projection of points on an object in space approximates the way we form a visual image of that object. The basic geometry of a perspective transformation includes the position of the observer, usually denoted as point O, the plane of projection (or picture plane), and λ, which is the normal distance from the observer to the plane. The simplest arrangement of these elements (Figure 9.28) places the observer on the z axis at $z = \lambda$, the distance from the xy plane, which we define as the picture plane.

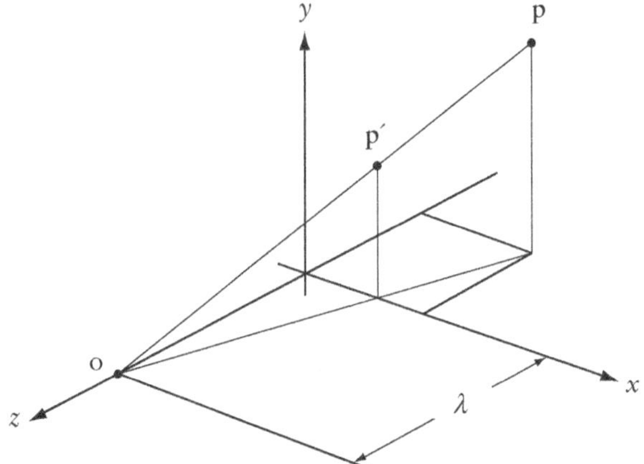

Figure 9.28 Central projection of points in space

Using the properties of similar triangles, we find

$$\frac{x'}{\lambda} = \frac{x}{\lambda - z}, \qquad \frac{y'}{\lambda} = \frac{y}{\lambda - z}, \qquad z' = 0, \tag{9.77}$$

or

$$x' = \frac{\lambda x}{\lambda - z}, \qquad y' = \frac{\lambda y}{\lambda - z}, \qquad z' = 0. \tag{9.78}$$

In matrix form these become

$$\begin{bmatrix} x' \\ y' \\ z' \\ 1 \end{bmatrix} = \begin{bmatrix} 1 & 0 & 0 & 0 \\ 0 & 1 & 0 & 0 \\ 0 & 0 & 0 & 0 \\ 0 & 0 & -1/\lambda & 1 \end{bmatrix} \begin{bmatrix} x \\ y \\ z \\ 1 \end{bmatrix}. \tag{9.79}$$

Here for the first time we use a position in the last row of the matrix (other than a_{44}, of course). It is important to note that in a left-hand coordinate system $a_{34} = 1/\lambda$. Try to verify this. If we cannot arrange the basic geometry as it is in Figure 9.28, then we may use two- or three-point perspective to achieve a realistic representation. Here the other elements in the last row of the transformation matrix, a_{41} and a_{42}, come into play, although we will not address these variations here.

We conclude with the vector solution to a general perspective projection of any point **p** onto an arbitrary plane relative to the center of projection $\mathbf{p}_0$ (Figure 9.29).

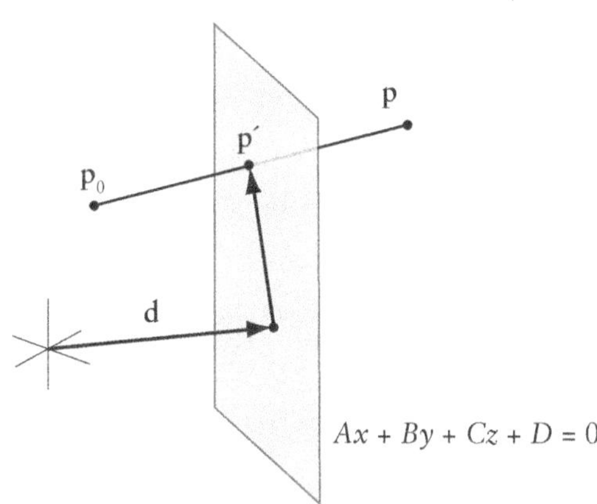

Figure 9.29 Vector solution for a general perspective projection

If the plane equation is $Ax + By + Cz + D = 0$, then

$$\mathbf{p}' = k(\mathbf{p} - \mathbf{p}_0) \tag{9.80}$$

and

$$(\mathbf{p}' - \mathbf{d}) \bullet \mathbf{d} = 0, \tag{9.81}$$

where from elementary geometry

$$d_x = \frac{AD}{A^2 + B^2 + C^2}, \tag{9.82}$$

$$d_y = \frac{BD}{A^2 + B^2 + C^2}, \tag{9.83}$$

$$d_z = \frac{CD}{A^2 + B^2 + C^2}. \tag{9.84}$$

Vector equations 9.80 and 9.81 produce four algebraic equations in four unknowns, k, p'_x, p'_y, and p'_z, which we easily solve to find $\mathbf{p}'$. The vector $\mathbf{d}$ is, of course, the normal vector to the plane from the origin. Equation 9.80 ensures that $\mathbf{p}'$ lies along the line joining $\mathbf{p}$ and $\mathbf{p}_0$, and Equation 9.81 ensures that $\mathbf{p}'$ lies on the projection plane.

9.8 Sweep Transformations

This section uses material from Geometric Modeling, *3rd edition, 2006. It serves as an introduction to an important part of 3D solid modeling, showing how vectors and transformations work together to produce a variety of shapes.*

Before we discuss the role of vectors and matrices in sweep transformations, let's get a general idea of just what they are. Sweep transformations are based on the notion of moving a curve, surface, or solid along some path. The locus of points generated by this process defines a new two- or three-dimensional object . . . a sweep surface. The relatively constrained geometry of a sweep shape means that only a small data set is needed to specify the shape.

We define a translational sweep by moving a planar curve segment or closed planar shape along a straight line normal to the plane of the curve, the former generating a surface and the latter a solid. We define a rotational sweep by rotating a planar curve or shape with finite area about an axis. Although this is a spatial rotation, the axis is usually in the plane of the figure . . . at least for the simplest and most common applications. Sweeping one curve along another curve, where the generating curve remains parallel to itself if it is a plane curve, produces a somewhat less restrictive definition of a translation surface.

A general sweep is one whose generating shape follows some arbitrary curved path, and which itself may change size, shape, and orientation. Thus, for modeling sweeps we need two primary ingredients: an object to move and a path to move it along. The sweeping object must be a curve, surface, or solid, and the path must be analytically definable. The term "generator" denotes the sweeping object, and the term "director" denotes the path. Unfortunately, the word "sweep," itself, is used as a noun, verb, or adjective. Fortunately, the meaning is usually clear from the context.

When does a sweep become so generalized that it is no longer a useful, intuitive concept in application? This is a difficult question, with no easy answer. However, we cannot go wrong if we insist on the presence of a generator shape, a director path, and rules controlling the orientation and shape evolution of the generator at each point on the director.

Sweep shapes are important in geometric modeling because they accurately represent a large class of engineering and manufacturing objects. In contemporary modeling systems, they prove to be practical and efficient for modeling constant cross-section mechanical parts. We may also use the swept-solid technique to detect potential interference between parts of mechanisms: For example, a moving object A collides with a fixed object B if the volume swept by A intersects B.

Figure 9.30 shows examples of sweep representations. The two principal types of trajectories are depicted: translation and rotation. Notice that the director curve is not necessarily an element of the swept object. Also, notice that the director of a rotational sweep amounts to an algorithm to move each point in the generator along a circular arc in a plane perpendicular to the axis of rotation and with a radius defined as the perpendicular distance from the point to the axis. In all cases, the shape of the object, or generator, being swept along does not change. Later we will investigate more general, nonlinear sweep representations.

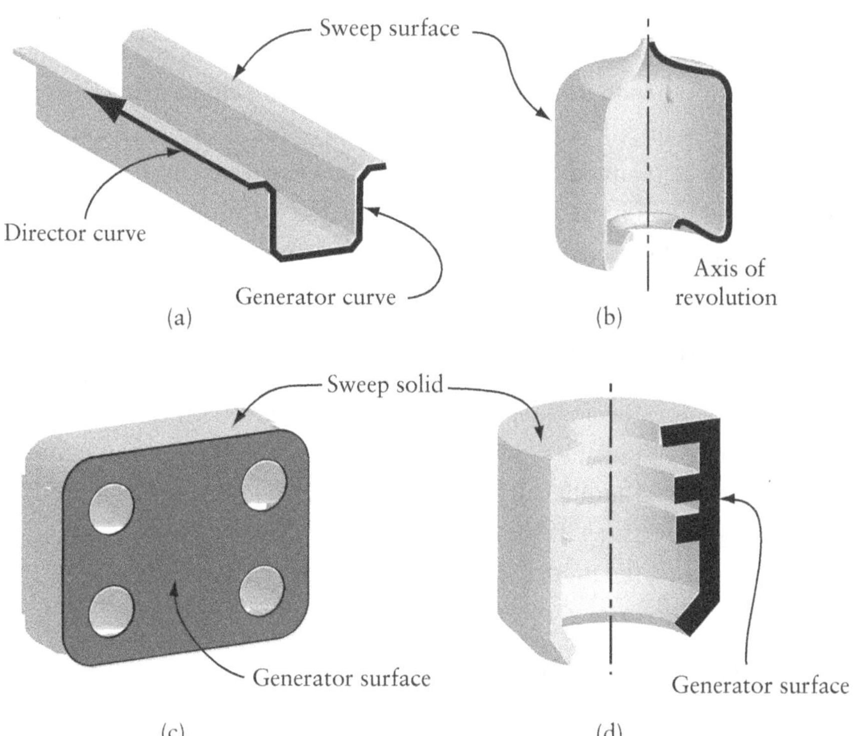

Figure 9.30 Examples of sweep shapes

There are several obvious and some not-so-obvious ways of creating dimensionally nonhomogeneous objects. In Figure 9.31a, the translational sweep of a curve creating a surface also creates two dangling edges. In Figure 9.31b, two two-dimensional regions are connected by a one-dimensional structure. In Figure 9.31c and 9.31d, creating solids using invalid or nonhomogeneous generators results in dimensionally nonhomogeneous solids and ambiguities.

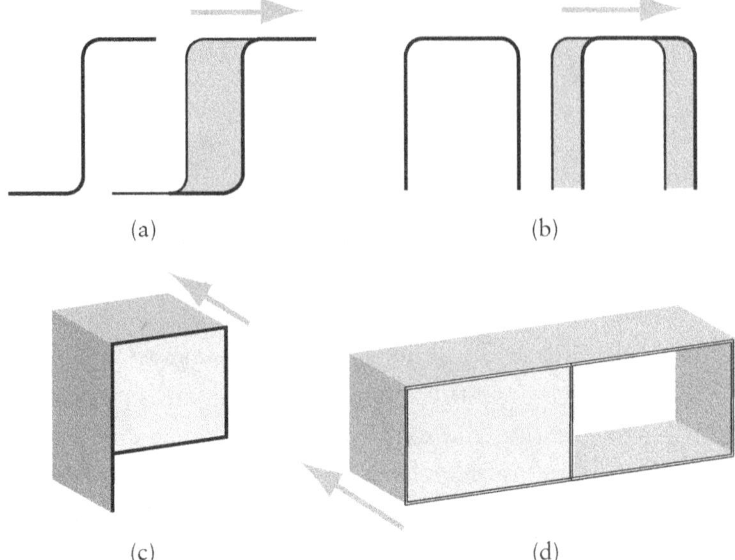

Figure 9.31 Dimensionally nonhomogeneous sweep representations

There are many other arrangements that produce similar problems: For example, the rotational sweep of a generator curve passing through the axis of rotation produces a surface or solid with a singularity. These conditions produce unacceptable results for most applications, but sometimes the results are as planned. By following criteria suggested in this figure and incorporating them in the design of sweep-representation-generating algorithms, we create dimensionally homogeneous models.

Translational Sweep

Figure 9.32 shows a closed planar curve defining the cross section of a model translated along a straight-line axis to form a surface. Limit planes at each end furnish bounding surfaces to complete a closed solid. Figure 9.32 illustrates these features. If this model followed a curved axis, a nonlinear transformation on the model would describe this representation.

Generalized Translational Sweep

A more generalized sweep requires a generator curve, as well as position and orientation curves (sometimes referred to as position-direction curves). With it we can define an almost unlimited variety of swept solids. A six-component position-direction curve, usually some form of parametric cubic curve, continuously specifies position and an associated direction or orientation. The first three components define position in three-dimensional space. The second three components define a corresponding continuous

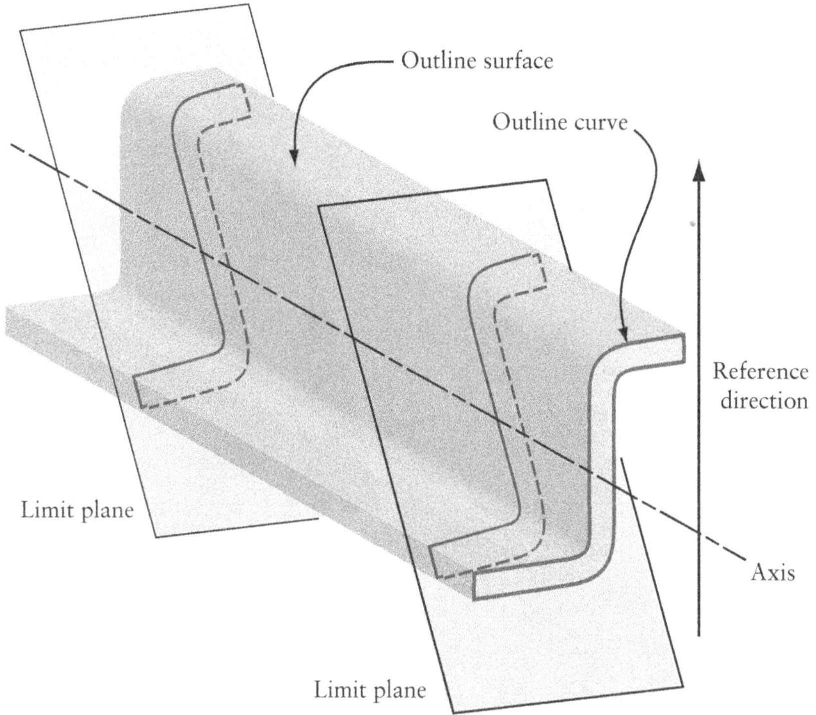

Figure 9.32 Outline surface of a constant cross-section solid

direction vector. A common parametric variable associates the direction vector with a specific position on the curve. Figure 9.33 shows the resulting curved and twisting coordinate system.

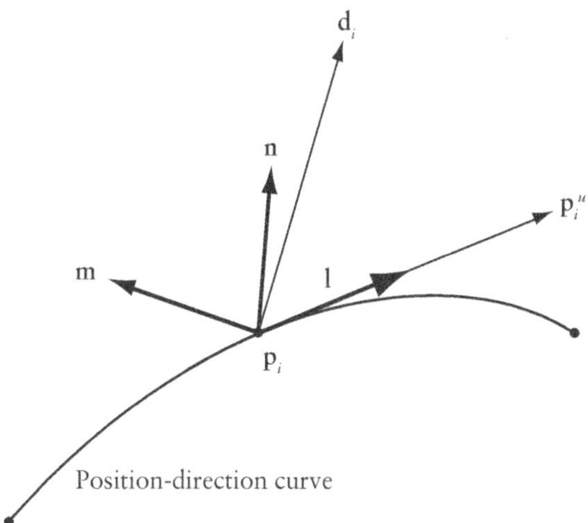

Figure 9.33 Characteristics of a position-direction curve

Here is how we interpret the position-direction curve. Construct a local orthogonal system at $\mathbf{p}_i$ as follows: Compute $\mathbf{p}_i^u$, the tangent vector. From $\mathbf{p}_i^u$ and $\mathbf{d}_i$ (where $\mathbf{d}_i$ is a direction associated with $\mathbf{p}_i$ and given by another parametric equation), we find the orthogonal unit vectors **l**, **m**, and **n**.

The axes **l** and **n** define a direction plane in which $\mathbf{d}_i$ is located. This reference plane changes as the tangent vector and $\mathbf{d}_i$ change continuously along the position-direction curve. A position-direction curve defines continuous transformations for points on an outline curve or generator. (Notice that a position-direction curve is essentially the equivalent of a moving trihedron.)

$$\mathbf{l} = \frac{\mathbf{p}_i^u}{|\mathbf{p}_i^u|}, \qquad \mathbf{m} = \frac{\mathbf{d}_i \times \mathbf{p}_i^u}{|\mathbf{d}_i \times \mathbf{p}_i^u|}, \qquad \mathbf{n} = \mathbf{l} \times \mathbf{m}. \tag{9.85}$$

We can use a position-direction curve to produce generalized outline surfaces. Figure 9.34 illustrates a constant cross-section part that not only curves but also twists.

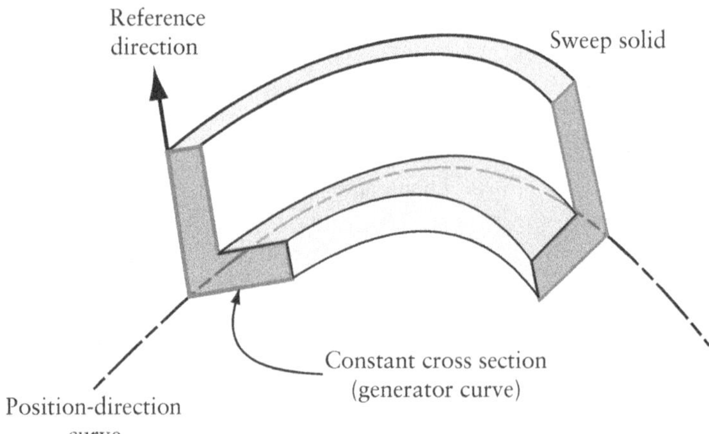

Figure 9.34 A constant cross-section part that curves and twists

Figure 9.35 illustrates how to coordinate the two curves that make up a position-direction curve through a common value of the parametric variable and how to extract the elements.

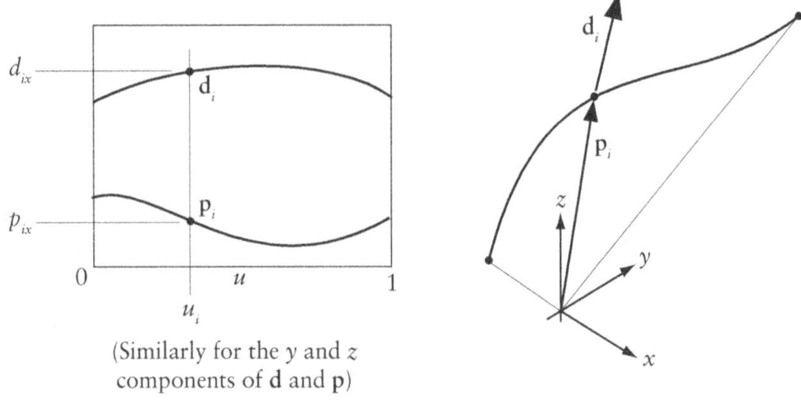

Figure 9.35 Components of a position-direction curve

Notice that more than one cross-section curve can be synchronized to one position-direction curve (for example, a pipe or tube with an inner and outer cross-section curve). Multiple cross sections with multiple position-direction curves are useful for generating a more complex or variable cross section. Positive and negative generator curves are used to create complex cross sections with high variability in the axial direction. Multiple position-direction curves and associated generator curves create, in effect, half-spaces with directed normals and can be used to add or subtract material to a raw-stock model. Some of these modeling procedures must depend on the user to verify model validity.

A six-component curve trajectory for controlling the movement of a generator curve has obvious natural extensions. For example, three more components may be added to yield scale factors to apply to the generator curve, differentially expanding or contracting it, and coordinating it with the motions imposed by the position-direction curve.

Rotational Sweep

One way to generate a surface of revolution is by revolving a plane curve around an axis line in its plane (Figure 9.36).

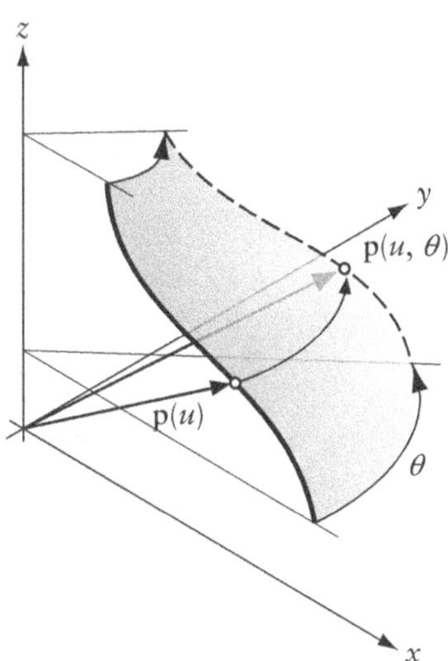

Figure 9.36 Surface of revolution

The plane curve is a profile curve, and in its various positions around the axis, it creates meridians. The circles created by each point on this curve are parallels. For the simplest case, we let the z axis be the axis of rotation and define the curve $\mathbf{p}(u) = \mathbf{x}(u) + \mathbf{z}(u)$ in the xz plane. Then the surface of revolution has the equation

$$\mathbf{p}(u,\theta) = x(u)\cos\theta\mathbf{i} + x(u)\sin\theta\mathbf{j} + z(u)\mathbf{k}, \tag{9.86}$$

where the profile curve might be a general curve, elliptic segment, circular arc, etc.

Translation with Deformation Sweep

Given an open or closed curve in the l, m, n Cartesian coordinate system, we can sweep the curve through space while scaling or otherwise deforming it to produce a surface or solid shape (Figure 9.37a). First, we define a parametric curve $\mathbf{p}(u)$ that describes the successive positions of the cross section, and define another parametric function $\mathbf{d}(u)$ that describes the directed orientation of the cross section. Three mutually orthogonal unit vectors $\mathbf{l}$, $\mathbf{m}$, and $\mathbf{n}$ give the orientation of the l, m, and n coordinate system at any point $\mathbf{p}(u)$, and we define these as

$$\mathbf{l} = \frac{\mathbf{p}^u}{|\mathbf{p}^u|}, \qquad \mathbf{m} = \frac{\mathbf{d} \times \mathbf{p}^u}{|\mathbf{d} \times \mathbf{p}^u|}, \qquad \mathbf{n} = \mathbf{l} \times \mathbf{m}, \tag{9.87}$$

where $\mathbf{p}^u = d\mathbf{p}/du$. Notice that $\mathbf{p}(u)$ and $\mathbf{d}(u)$ are coordinated through their common parametric variable u. The vector function $\mathbf{d}(u)$ determines the rotational orientation of $\mathbf{m}$ and $\mathbf{n}$ about an axis coincident with $\mathbf{l}$. Equation 9.87 ensures that $\mathbf{l}$, $\mathbf{m}$, and $\mathbf{n}$ do indeed form a mutually orthogonal triad (Figure 9.37b).

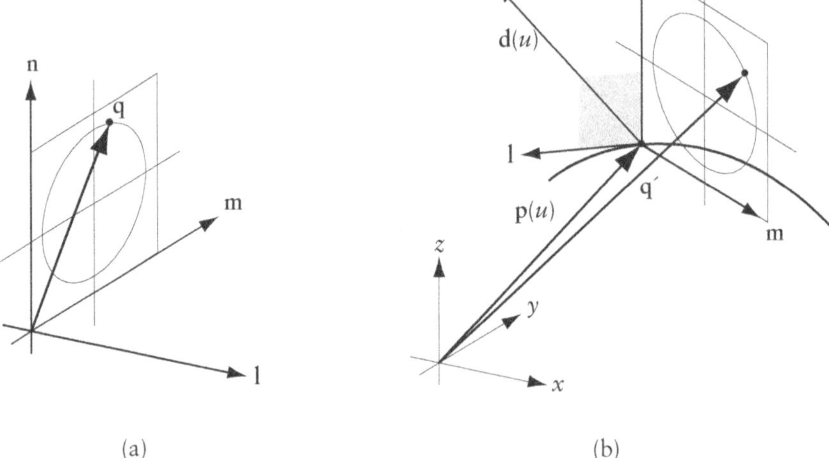

Figure 9.37 Sweep transformation and deformation

Thus, the sweep transformation of any point $\mathbf{q}$ in the initial l, m, and n coordinate system onto the x, y, and z system at $\mathbf{p}(u)$ is $\mathbf{q}'$, where

$$\mathbf{q}' = \mathbf{p}(u) + q_l\mathbf{l} + q_m\mathbf{m} + q_n\mathbf{n}. \tag{9.88}$$

This relationship reconstructs or maps the initial curve in l, m, and n space at any point $\mathbf{p}(u)$ in x, y, z space. If the initial curve is one we define by a set of control points, then these points are swept into new positions according to Equation 9.88. The complete curve is then regenerated based on the transformed control points.

We may also continuously change the scale of the curve as it sweeps along $\mathbf{p}(u)$ by simply adding another parametric function, say $\mathbf{s}(u)$, constructed to yield the appropriate scale factors (rescaled in the l, m, n system). Now all these functions $\mathbf{p}(u)$, $\mathbf{d}(u)$, and $\mathbf{s}(u)$ operate in a coordinated fashion to position, direct, and scale the initial curve. Since $\mathbf{s}(u)$ is also a vector function, it can be used to produce three separate scale factors s_ℓ, s_m, s_n.

9.9 Exercises

9.1. Given a homogeneous transformation **H**, partition this matrix and show that the 3×3 submatrix in the upper-left partition is always orthogonal.

9.2. Show that the determinant of the translation transformation in homogeneous coordinates is equal to 1.

9.3. Sketch the central inversion of a rectangle in the plane, and discuss the orientation of the original and inverted rectangle.

9.4. Find the matrix and Cartesian equations that describe a reflection fixing the line $x = d$ parallel to the y axis.

9.5. Find the reflection matrix that maps any point on the plane onto its mirror image across the line $y = x / \sqrt{3}$.

9.6. Show that the product of the reflection matrices $\begin{bmatrix} 0 & 1 & 0 \\ 1 & 0 & 0 \\ 0 & 0 & 1 \end{bmatrix}$ and $\begin{bmatrix} 0 & -1 & 0 \\ -1 & 0 & 0 \\ 0 & 0 & 1 \end{bmatrix}$ is a rotation. Describe the rotation.

9.7. Show that any polygon transforms into a similar polygon under a scaling transformation.

9.8. Show that angle size is preserved under a scaling s.

9.9. Find the center of scaling for the transformation given by the matrix

$$\begin{bmatrix} s & 0 & c_1 \\ 0 & s & c_2 \\ 0 & 0 & 1 \end{bmatrix}.$$

9.10. Show that $\mathbf{SS}^{-1} = 1$.

9.11. Write the Cartesian equations for a shear transformation in the plane that fixes the y axis.

10 Vector-Defined Geometric Objects I

Vectors, plain and simple, and vector functions are efficient ways to represent and analyze geometric objects, from points, lines, and planes to complex curves and surfaces. That is just the beginning . . . Using vector algebra, we can compute their local and global geometric properties, as well as spatial relationships such as distances and intersections.
This chapter introduces some of the ways to define and analyze points, lines, planes, polygons, and polyhedra. The next chapter moves on to curves and surfaces.

10.1 Points

A vector may represent a point. Its components are the coordinates of the point. Every point useful to geometric modeling has a vector associated with it. And from this all the other elements follow. Here is a closer look at a point.

The point is the basic building block for all other geometric objects. It is the simplest of the three elementary forms: points, lines, and planes. It cannot be defined in terms of anything simpler. Elementary geometry demonstrates how geometric figures are defined as a locus of points, each figure with certain unique constraining properties. For example, remember that in a plane, a circle is the locus of points equidistant from a given point, and a straight line is the locus of points equidistant from two given points.

There are simple equations that are the equivalent of those locus statements. In three-dimensional space, a plane is the locus of points equidistant from two given points. This process applies to even more complex objects, for example, curves, surfaces, and solids. To do this, sets of equations provide the constraints that define the locus of points that define the geometric object. It is a powerful way to describe these objects, because these equations or functions allow analysis and quantification of their properties and relationships. Points provide the controls for shaping, moving, and transforming geometric objects, that is, points acting as vectors.

There is (usually) a one-to-one correspondence between the coordinates of a point and the components of an associated vector. We will continue to use a boldface, lowercase $\mathbf{p}$ to represent its vector format and a boldface uppercase **P** to represent the matrix form of a point. To review, here is point representation in matrix form:

$$\mathbf{P} = [\ x\mathbf{i} \quad y\mathbf{j} \quad z\mathbf{k}\ ], \tag{10.1}$$

or alternatively,

$$\mathbf{P} = [\ p_x\mathbf{i} \quad p_y\mathbf{j} \quad p_z\mathbf{k}\ ]. \tag{10.2}$$

Dropping the attached orthogonal unit vectors produces

$$\mathbf{P} = [\ x \quad y \quad z\ ], \tag{10.3}$$

or

$$\mathbf{P} = [\ p_x \quad p_y \quad p_z\ ]. \tag{10.4}$$

For our purposes, these are all equivalent forms, with

$$\begin{aligned} x &= p_x, \\ y &= p_y, \\ z &= p_z. \end{aligned} \tag{10.5}$$

When many points are in play, subscripts identify individual points. So for point vector $\mathbf{p}_i$ we have in matrix form:

$$\mathbf{P}_i = [\ x_i \quad y_i \quad z_i\ ]. \tag{10.6}$$

10.2 Lines

Lines are the next simplest geometric objects. Here are two ways to use vectors to define a line in space.

Line Through a Point and Parallel to a Vector

The vector equation of a line through some point $\mathbf{p}_0$ and parallel to another vector **t** is

$$\mathbf{p}(u) = \mathbf{p}_0 + u\mathbf{t}, \tag{10.7}$$

where u is a scalar independent variable multiplying **t** (Figure 10.1). The vector **t** is usually a unit vector, but not necessarily so.

As u takes on successive values, the equation generates points on a straight line. The components of $\mathbf{p}(u)_i$ are the coordinates of a point on this line. In other words, because $\mathbf{p}_0$ and **t** are constant and define a specific line, any real value of u generates a point on that line.

Now let's see what happens when we expand the vector equation $\mathbf{p}(u) = \mathbf{p}_0 + u\mathbf{t}$ by writing it in its component form and listing the components

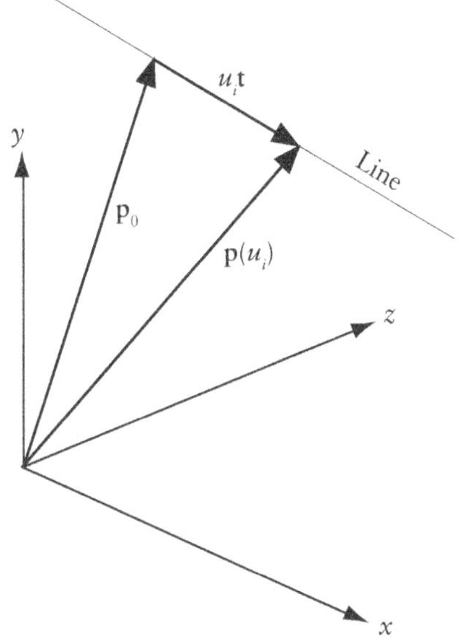

Figure 10.1 Vector definition of a line. (A left-hand coordinate system is shown here.)

in a column matrix instead of the horizontal or row array used earlier. (The row and column forms are mathematically equivalent, demanding only that we do not mix the two and that we use some simple bookkeeping techniques when doing algebra on them.) This produces

$$\begin{bmatrix} x \\ y \\ z \end{bmatrix} = \begin{bmatrix} x_0 \\ y_0 \\ z_0 \end{bmatrix} + u \begin{bmatrix} t_x \\ t_y \\ t_z \end{bmatrix}. \tag{10.8}$$

In polynomial form this matrix equation produces the following three linear equations:

$$\begin{aligned} x &= x_0 + ut_x, \\ y &= y_0 + ut_y, \\ z &= z_0 + ut_z, \end{aligned} \tag{10.9}$$

where u is the independent variable; x, y, and z are dependent variables; and x_0, y_0, z_0, t_x, t_y, and t_z are constants. Mathematicians call this set of equations the parametric equations of a straight line: (x, y, z) are the coordinates of any point on the line, (x_0, y_0, z_0) are the coordinates of a given point on the line, and (t_x, t_y, t_z) are the components of a vector parallel to the line.

Here is a simple example in two dimensions, lying in the xy plane (Figure 10.2): Let's find the vector equation for a straight line that passes through the point described by the vector $\mathbf{p}_0 = (1,4)$ and parallel to the vector $\mathbf{t} = (2, -1)$.

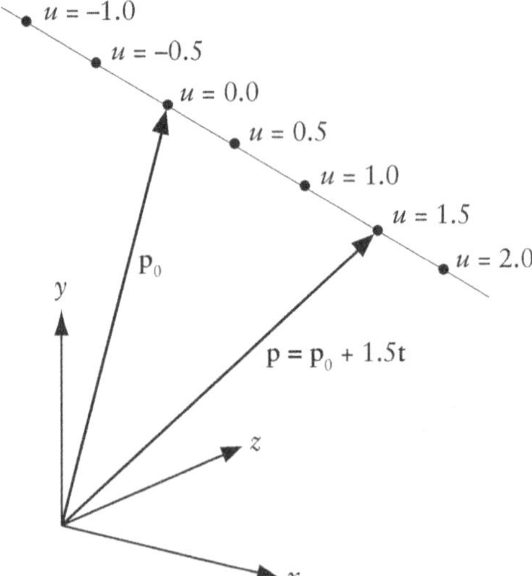

Figure 10.2 Example of points on a line

The matrix equation for this line in the *xy* plane is

$$\mathrm{P} = \begin{bmatrix} x \\ y \end{bmatrix} = \begin{bmatrix} 1 \\ 4 \end{bmatrix} + u \begin{bmatrix} 2 \\ -1 \end{bmatrix}, \tag{10.10}$$

or in parametric polynomial form:

$$\begin{aligned} x &= 1 + 2u, \\ y &= 4 - u. \end{aligned} \tag{10.11}$$

Considering three dimensions, Equation 10.8 is

$$\mathrm{P} = \begin{bmatrix} x \\ y \\ z \end{bmatrix} = \begin{bmatrix} 1 \\ 4 \\ 0 \end{bmatrix} + u \begin{bmatrix} 2 \\ -1 \\ 0 \end{bmatrix}. \tag{10.12}$$

Now compute the coordinates of points on this line for a series of values of *u* and list the results; for example, see the table below.

u	x	y	z
−1.0	−1.0	5.0	0
−0.5	0	4.5	0
0	1.0	4.0	0
0.5	2.0	3.5	0
1.0	3.0	3.0	0
1.5	4.0	2.5	0
2.0	5.0	2.0	0

Of course, we can easily expand this listing in a variety of ways. For example, we can use values of *u* closer together or farther apart, as well as values outside the range shown here.

Line Defined by Two Points

We can make a small change to Equation 10.7 that will let us find the vector equation of a line through two given points, say, $\mathbf{p}_0$ and $\mathbf{p}_1$ (Figure 10.3). In this figure, we see that we can define $\mathbf{t}$ as $\mathbf{p}_1 - \mathbf{p}_0$.

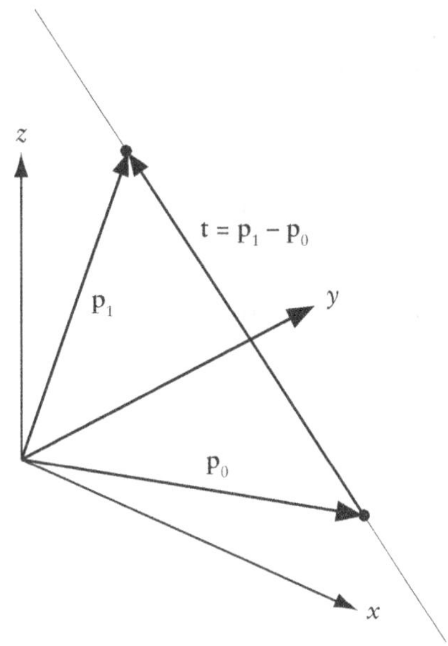

Figure 10.3 Vector definition of a line through two points

Making the appropriate substitution into Equation 10.7, we obtain

$$\mathbf{p} = \mathbf{p}_0 + u(\mathbf{p}_1 - \mathbf{p}_0), \tag{10.13}$$

which is just a compact form of the three polynomials:

$$\begin{aligned} x &= x_0 + u(x_1 - x_0), \\ y &= y_0 + u(y_1 - y_0), \\ z &= z_0 + u(z_1 - z_0). \end{aligned} \tag{10.14a}$$

By restricting the values of u to the interval $0 \le u \le 1$, this equation defines a line segment extending from $\mathbf{p}_0$ to $\mathbf{p}_1$. Using Equation 10.14a, let's see why this is true.

For $u = 0$,

$$\begin{aligned} x &= x_0, \\ y &= y_0, \\ z &= z_0. \end{aligned} \tag{10.14b}$$

For $u = 1$,

$$\begin{aligned} x &= x_0 + u(x_1 - x_0) = x_0 + (x_1 - x_0) = x_1, \\ y &= y_0 + u(y_1 - y_0) = y_0 + (y_1 - y_0) = y_1, \\ z &= z_0 + u(z_1 - z_0) = z_0 + (z_1 - z_0) = z_1. \end{aligned} \tag{10.14c}$$

So we see that this equation returns $\mathbf{p}_0$ and $\mathbf{p}_1$ for parametric variable values of $u = 0$ and $u = 1$, respectively.

10.3 Planes

Planes are indispensable to geometric and 3D modeling. The polygonal faces that often approximate curved surfaces lie in planes and are key to efficient computation of intersections, visibility, and rendering. Computer graphics images are projected onto planes. Planes intersect models to reveal cross-section geometry. Here are four ways to define a plane using vector equations.

Through a Point and Parallel to Two Independent Vectors

First, let's look at the vector equation of a plane through a point $\mathbf{p}_0$ and parallel to two independent vectors **s** and **t**. We denote the vector as $\mathbf{p}(u, w)$, such that

$$\mathbf{p}(u, w) = \mathbf{p}_0 + u\mathbf{s} + w\mathbf{t}, \tag{10.15}$$

where $\mathbf{p}(u, w)$ depends on the parametric variables u and w defining a point on the plane and where $\mathbf{s} \neq k\mathbf{t}$ (Figure 10.4).

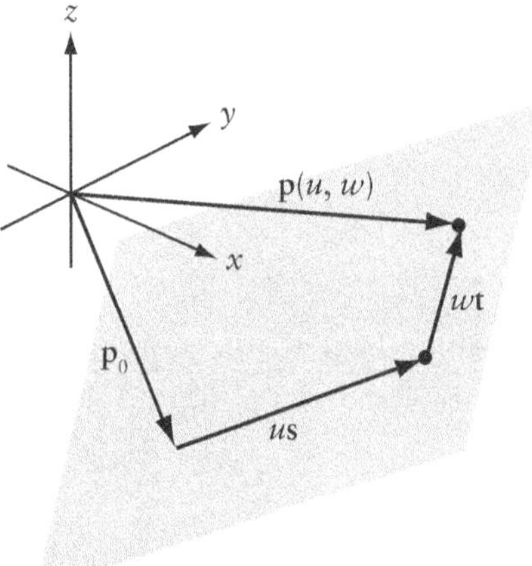

Figure 10.4 Vector definition of a plane

The vector $\mathbf{p}(u, w)$ represents the set of points defining a plane as the parameters u and w vary independently. In terms of the vector's components, the dependent variables x, y, and z are

$$\begin{aligned} x &= x_0 + us_x + wt_y, \\ y &= y_0 + us_y + wt_y, \\ z &= z_0 + us_z + wt_z, \end{aligned} \tag{10.16}$$

or as the matrix equation

$$\begin{bmatrix} x \\ y \\ z \end{bmatrix} = \begin{bmatrix} x_0 \\ y_0 \\ z_0 \end{bmatrix} + u \begin{bmatrix} s_x \\ s_y \\ s_z \end{bmatrix} + w \begin{bmatrix} t_x \\ t_y \\ t_z \end{bmatrix}. \tag{10.17}$$

Three-Point Definition

Three noncollinear points $\mathbf{p}_0$, $\mathbf{p}_1$, and $\mathbf{p}_2$ are sufficient to define a plane in space. Rewriting Equation 10.15 in terms of these points produces

$$\mathbf{p}(u,w) = \mathbf{p}_0 + u(\mathbf{p}_1 - \mathbf{p}_0) + w(\mathbf{p}_2 - \mathbf{p}_1). \tag{10.18}$$

We'll denote a vector that is normal to a plane as $\mathbf{n}$. Here are two ways to compute it. First, given two noncollinear vectors lying in the plane, $\mathbf{s}$ and $\mathbf{t}$ (for example, see Figure 10.4),

$$\mathbf{n} = \mathbf{s} \times \mathbf{t}, \tag{10.19}$$

or using the three points $\mathbf{p}_0$, $\mathbf{p}_1$, and $\mathbf{p}_2$, where $\mathbf{s} = \mathbf{p}_1 - \mathbf{p}_0$ and $\mathbf{t} = \mathbf{p}_2 - \mathbf{p}_1$, we have

$$\mathbf{n} = (\mathbf{p}_1 - \mathbf{p}_0) \times (\mathbf{p}_2 - \mathbf{p}_1). \tag{10.20}$$

Notice that we can construct a normal at any point on the plane and that, of course, all normals to the plane are parallel to one another. If the magnitude of $\mathbf{n}$ is not of interest, we can work with the unit normal $\hat{\mathbf{n}}$,

$$\hat{\mathbf{n}} = \frac{\mathbf{n}}{|\mathbf{n}|}. \tag{10.21}$$

Three points or vectors also define the vertices of a three-sided polygon. Recall from school geometry that two points define a line in space, and three points a plane. These, in turn, are the defining elements of polygons and polyhedra. Point vectors become the vertices. Lines become their edges, and planes their faces. Sets of polygons are used in geometric modeling to approximate the curved boundary surface of a solid. A graphics program can then project, fill, and shade these polygons to create realistic images of 3D solids.

Point on the Plane and the Unit Normal to It

A third way to define a plane is by using a point it passes through, $\mathbf{p}_0$, and the normal vector to it, $\mathbf{n}$. This implies that any point $\mathbf{p}$ lies on the plane if and only if $\mathbf{p} - \mathbf{p}_0$ is perpendicular to $\hat{\mathbf{n}}$, because $\hat{\mathbf{n}}$ is perpendicular to all lines in the plane. In terms of a vector equation, this statement becomes

$$(\mathbf{p} - \mathbf{p}_0) \bullet \hat{\mathbf{n}} = 0. \tag{10.22}$$

Remember: The scalar product of two mutually perpendicular vectors is zero. Performing the indicated scalar product yields

$$(x-x_0)\hat{n}_x+(y-y_0)\hat{n}_y+(z-z_0)\hat{n}_z=0, \tag{10.23a}$$

where $\hat{n}_x$, $\hat{n}_y$, and $\hat{n}_z$ are the components of $\hat{\mathbf{n}}$. If we rewrite this as

$$x\hat{n}_x+y\hat{n}_y+z\hat{n}_z+(-x_0\hat{n}_x-y_0\hat{n}_y-z_0\hat{n}_z)=0, \tag{10.23b}$$

we see that this is equivalent to the familiar equation of a plane

$$Ax+By+Cz+D=0. \tag{10.23c}$$

Normal Vector from the Origin to the Plane

The fourth way to define a plane is a variation of the third way: Given the vector **d** to a point on the plane, where **d** itself is perpendicular to the plane, then any point **p** on the plane must satisfy (Figure 10.5)

$$(\mathbf{p}-\mathbf{d})\bullet\mathbf{d}=0. \tag{10.24}$$

In this case, we should assume that **d** is a fixed vector; that is, its tail is fixed at the origin.

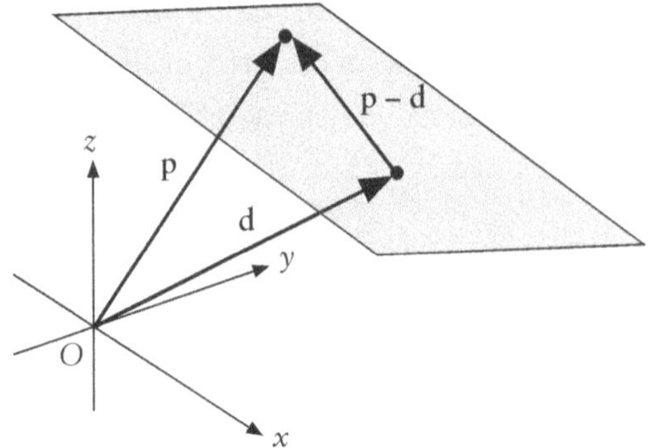

Figure 10.5 Normal vector from origin to a plane

10.4 Polygons

Polygons are surprisingly simple arrangements of points, lines, and planes. They are a big part of geometric 3D modeling, computer graphics, and other applications because it is easy to approximate the bounding surfaces of solids with a set of connected polygons. Rendering programs then project, fill, and shade each polygon to create realistic images of 3D solids.

Definition of a Polygon

A polygon is a many-sided two-dimensional figure bounded by edges (straight-line segments) and vertices (corner points). We will address polygons whose vertices all lie in the same plane. Figure 10.6a–d shows examples of polygons with three, four, five, and six sides, respectively. There is no limit to the number of edges a polygon can have. In modeling, triangles are by far the most used.

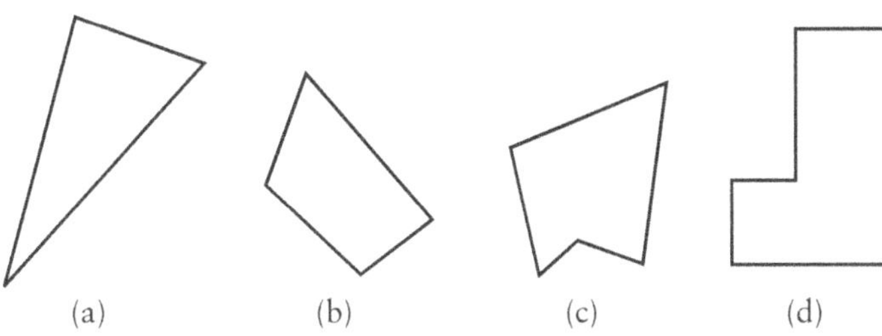

Figure 10.6 A variety of polygons

A plane polygon is convex, concave, or stellar. If the straight lines that are prolongations of the bounding edges of a polygon do not penetrate the interior, and if its edges intersect only at vertices, then the polygon is convex; otherwise it is concave. For any concave polygon we can always find a convex one that contains it. Polygons a and b in Figure 10.6 are convex; polygons c and d are concave.

The most common polygons are the so-called regular polygons. A regular polygon lies in a plane, has straight-line edges all of equal length, has equal vertex angles, and can be inscribed in a circle with which it shares a common geometric center. Figure 10.7 shows several regular polygons. It is easy to see that there are an infinite number of regular polygons.

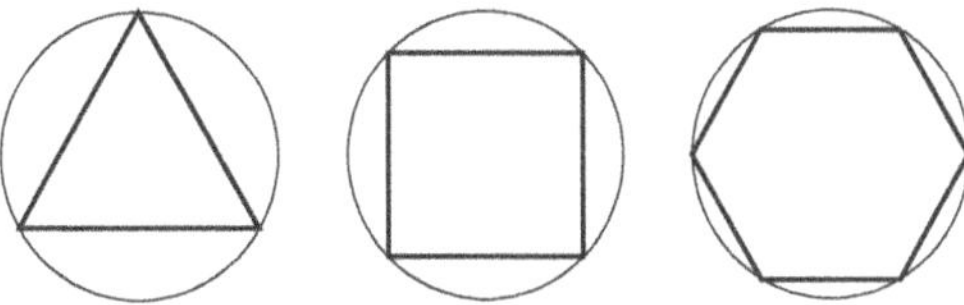

Figure 10.7 Regular polygons

A polygon is equilateral if all its sides are equal, and equiangular if all its angles are equal. If the number of edges is greater than three, then it can be equilateral without being equiangular (and vice versa). A rhombus is equilateral; a rectangle is equiangular.

Properties of Polygons

A plane polygon must have at least three edges to enclose a finite area. Notice that for all plane polygons, the number of vertices equals the number of edges, so that

$$V - E = 0, \tag{10.25}$$

where V is the number of vertices and E is the number of edges.

We can divide a plane polygon into a set of triangles (Figure 10.8) by a procedure called triangulation. As the figure shows, usually there is more than one way to do this. The minimum number of triangles T, or the minimum triangulation, is given by

$$T = V - 2. \tag{10.26}$$

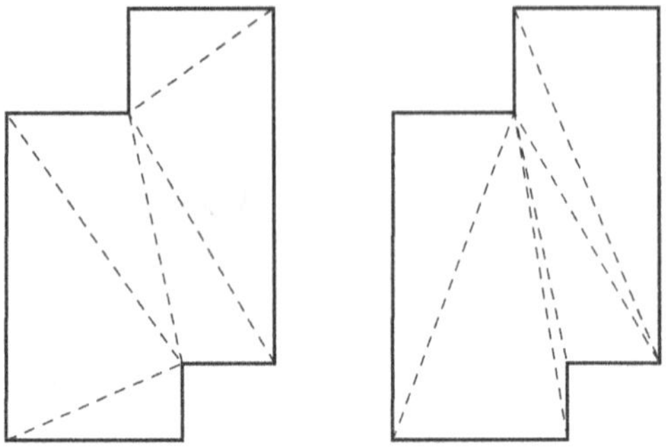

Figure 10.8 Different triangulations of a plane polygon

The sum of the exterior angles of a plane polygon is 2π. This means that each exterior angle of a regular polygon is $2\pi / E$ and its interior angle (the supplement) is $(1 - 2/E)\,\pi$ (Figure 10.9).

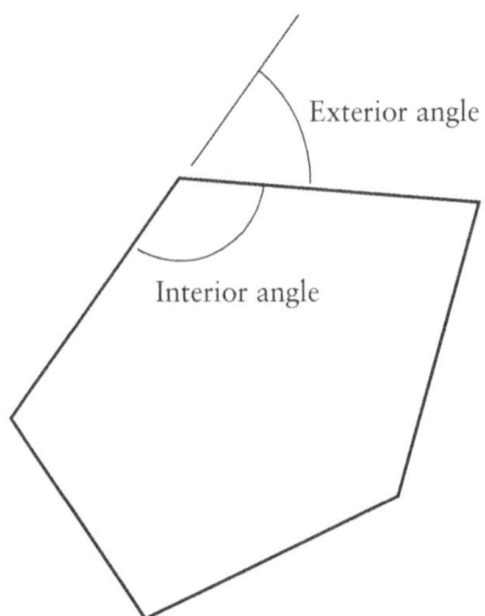

Figure 10.9 Exterior and interior angles

The sum of the interior angles of a polygon with E edges is

$$\sum_{V} \theta = 180(E-2). \tag{10.27}$$

An easy way to visualize this result is to triangulate the polygon with line segments radiating from an interior point to each of the vertices.

The average angle of a polygon is the sum of the angles divided by the number of angles, which equals E; that is,

$$\theta_{avg} = \frac{180(E-2)}{E}. \tag{10.28}$$

For regular polygons with E edges, the interior angles are identical (as are the edge lengths), so each interior angle equals the average value.

The perimeter of a regular polygon is

$$\text{Perimeter} = EL, \tag{10.29}$$

where E is the number of edges and L is the length of an edge.

The area of a regular polygon is

$$A = \frac{EL^2}{4}\cot\left(\frac{\pi}{E}\right). \tag{10.30}$$

We can describe a polygon by simply listing the coordinates of its vertex points: $x_1, y_1, x_2, y_2, \ldots, x_n, y_n$ or $\mathbf{p}_1, \mathbf{p}_2, \ldots \mathbf{p}_n$, as vectors, where n is the number of vertices. Successive vertex points define a polygon's edges, where the last edge is the line between x_n, y_n and x_1, y_1. Using such a list, we find that the average x and y values of the coordinates of its vertices define the geometric center, gc, of a convex polygon; thus,

$$\begin{aligned} x_{gc} &= \frac{x_1 + x_2 + \ldots + x_n}{n}, \\ y_{gc} &= \frac{y_1 + y_2 + \ldots + y_n}{n}. \end{aligned} \tag{10.31}$$

Vector geometry provides an easy way to find the area of a polygon. First, to compute the area of a triangle, select a reference vertex $\mathbf{p}_1$. The area is one-half the absolute value of a vector product:

$$A = \frac{1}{2}\left|(\mathbf{p}_2 - \mathbf{p}_1)\times(\mathbf{p}_3 - \mathbf{p}_1)\right|, \tag{10.32}$$

where position vectors give the three vertices of the triangle $\mathbf{p}_i$.

To compute the area of a polygon with four or more edges, select a reference vertex $\mathbf{p}_1$. The area of a plane polygon with n vertices, where $n \geq 4$, is

$$A = \frac{1}{2}\left|\sum_{i=1}^{n-2}(\mathbf{p}_{i+1} - \mathbf{p}_1)\times(\mathbf{p}_{i+2} - \mathbf{p}_1)\right|. \tag{10.33}$$

Note that, in general, there are n – 2 "signed" (oriented) triangular areas. The vector product nicely accounts for triangle overlaps and, thus, holds for either convex or concave polygons.

Convex Hull of a Polygon

The convex hull of any polygon is the convex polygon that is formed by stretching a rubber band over the vertex points (Figure 10.10). The convex hull of a convex polygon has edges corresponding identically to those of the polygon itself, while the convex hull of a concave polygon always has fewer edges than the concave polygon itself, some edges of which are not collinear, and it encloses a larger area.

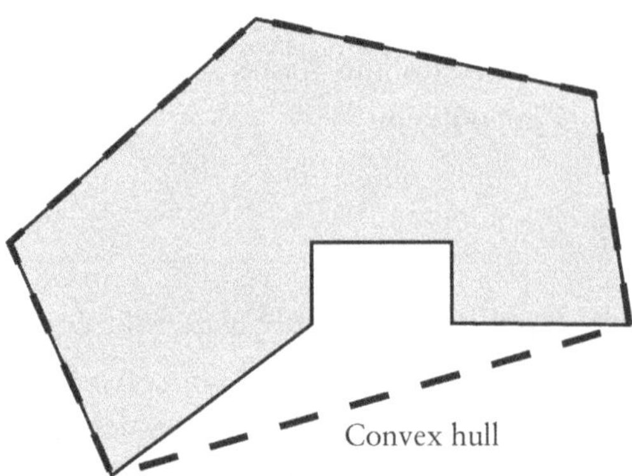

Figure 10.10 Convex hull of a concave polygon

A convex hull is also associated with any arbitrary set of points in the plane. Polyhedra and sets of points in space have three-dimensional convex hulls, analogous to the two-dimensional case. In fact, the convex hull of the points defining a Bézier curve or surface approximates the curve or surface, and it is useful in analyzing and modifying its shape, as we shall see.

Triangles: A Very Special Polygon

While efficient in the modeling phase, parametric bicubic bounding surfaces of geometric models and 3D solids pose some computational challenges when it comes time to render them into a computer-generated image. The ray-tracing phase of image rendering calls for finding intersections of lines (rays) and surfaces. The simplest and fastest intersection computations are those between lines and planes. This is where triangles come in; they are the simplest kind of bounded plane. Computing these intersections means finding the intersection point of the ray and plane and determining if this point lies within the triangle. It is easy to extract another important piece of information: the angle between the ray and the plane. *Hint:* Find the vector normal to the plane, then find the scalar product of it with the ray vector, and finally use Equation 3.35 to find the angle.

Data Structure for Polygons

The simplest data structure for a polygon is an $n \times 3$ matrix of its vertex coordinates v_i, $i = 1, n$, where n = number of vertex points. This scheme assumes that successive points are connected by a polygon edge and that the first and last vertices are connected:

$$\begin{bmatrix} x_1 & y_1 & z_1 \\ \vdots & \vdots & \vdots \\ x_i & y_i & z_i \\ \vdots & \vdots & \vdots \\ x_n & y_n & z_n \end{bmatrix}.$$

10.5 Polyhedra

We use combinations and assemblies of polyhedra, like building blocks, to create geometric models for CAD/CAM, computer graphics, and similar applications. Alone and in combinations, their properties are easy to compute. And what is most amazing, we can define all possible polyhedra by only the three simplest of geometric elements—points, lines, and planes, all of which we can represent with vectors.

Definition of a Polyhedron

A polyhedron is a multifaceted three-dimensional solid bounded by a finite, connected set of plane polygons such that every edge of each polygon belongs also to just one other polygon. The polygonal faces form a closed surface, dividing space into two regions, the interior and exterior of the polyhedron. A cube (or hexahedron) is a familiar example of a polyhedron.

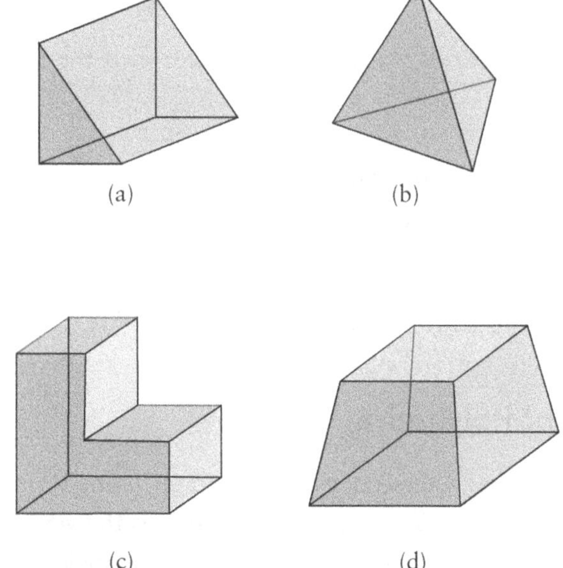

Figure 10.11 Examples of polyhedra

All the faces of a polyhedron are plane polygons, and all its edges are straight-line segments. The simplest polyhedron, one with four faces, is the tetrahedron (Figure 10.11b). Polyhedra a, b, and d in Figure 10.11 are convex, and polyhedron c is concave. In every case, exactly two polygonal faces share each polyhedral edge.

Three geometric elements define all polyhedra: vertices (V), edges (E), and faces (F). An equal number of edges and faces surrounds each vertex, two vertices and two faces bound each edge, and a closed loop of coplanar edges that form a polygon bounds each face. Finally, the angle between faces that intersect at a common edge is the dihedral angle. Any straight line in a plane divides the plane into two half-planes, and two half-planes extending from a common line form a dihedral angle. The cube in Figure 10.12 illustrates these elements of a polyhedron.

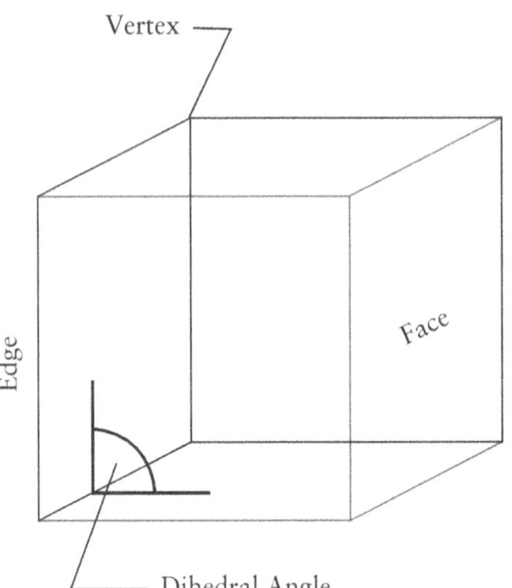

Figure 10.12 Dihedral angle

A convex polyhedron is a regular polyhedron if the following conditions are true:

- All face polygons are regular.
- All face polygons are congruent.
- All vertices are identical.
- All dihedral angles are equal.

Properties of Polyhedra

Three or more planes intersecting at a common point form a solid angle called a polyhedral angle, and the common point is the vertex of this angle. The intersections of the planes are the edges of the angle. The parts of the planes lying between the edges are the faces of the angle. The angles formed by adjacent edges are the face angles of the polyhedral angle. Thus, the vertex, edges, faces, face angles, and dihedral angles formed by the faces are the constituent parts of the polyhedral angle. For any polyhedral angle, there is an equal number of edges, faces, face angles, and dihedral angles. A polyhedral angle with three faces is a trihedral angle (the polyhedral angle at a vertex of a cube is an example). Polyhedral angles of four, five, six, and seven faces are called tetrahedral, pentahedral, hexahedral, and heptahedral, respectively.

An angle of 360° surrounds any point on a plane. The "angular deficit" at a vertex of a polyhedron is defined as the difference between the sum of the face angles surrounding the vertex and 360°. It is the gap that results if the solid angle at a vertex is opened out flat. The sum of the angular deficits over all the vertices of a polyhedron is its total angular deficit.

The sum of all face angles at a vertex of a convex polygon is always less than 2π. Otherwise, one of two conditions is present: If the sum of the angles is exactly 2π, then the edges meeting at the vertex are coplanar; if the sum of the angles is greater than 2π, then some of the edges at the vertex are reentrant and the polyhedron is concave.

Convex Hull of a Polyhedron

The convex hull of a polyhedron is the three-dimensional analog of the convex hull for a polygon. The convex hull of a convex polyhedron is identical to the polyhedron itself. We form the convex hull of a concave polyhedron by wrapping it in a rubber sheet, producing an enveloping convex polyhedron. The concave polyhedron in Figure 10.13a, for example, has the convex hull shown in Figure 10.13b. The convex hull of a concave polyhedron is the smallest convex polyhedron that will enclose it.

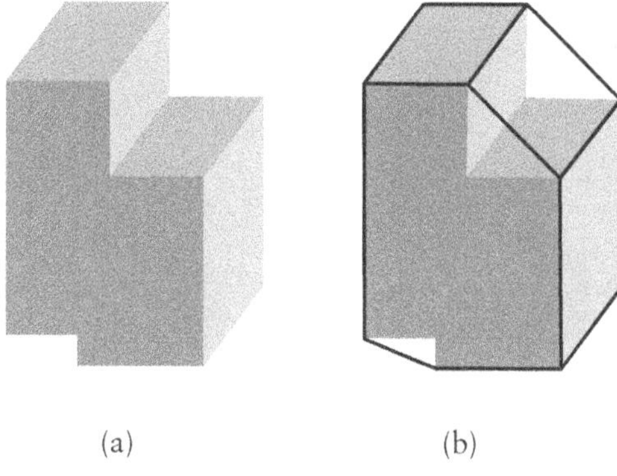

Figure 10.13 Convex hull of a concave polyhedron

Connectivity Matrix

We can efficiently organize the data describing a polyhedron by listing the vertices and their coordinates and by using a connectivity matrix to define its edges. It is a two-dimensional list or table that describes how vertices are connected by edges to form a polyhedron. This matrix is always square, with as many rows and columns as vertices. We denote a particular element or entry in the table as a_{ij}, where the subscript i tells us which row the element is in and j tells us which column. If element $a_{ij} = 1$, then vertices i and j are connected by an edge. If element $a_{ij} = 0$, then vertices i and j are not connected. A connectivity matrix is symmetric about its main diagonal, which is composed of all zeros. For example, the tetrahedron shown in Figure 10.14 has the following connectivity matrix:

$$\mathbf{C}_{\text{tetrahedron}} = \begin{bmatrix} 0 & 1 & 1 & 1 \\ 1 & 0 & 1 & 1 \\ 1 & 1 & 0 & 1 \\ 1 & 1 & 1 & 0 \end{bmatrix}. \tag{10.34}$$

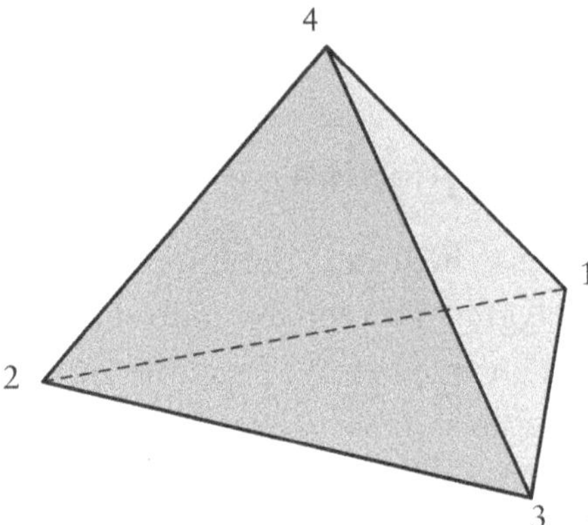

Figure 10.14 Tetrahedron connectivity

The columns are numbered consecutively from left to right and the rows from top to bottom. This method readily lends itself to the design of a database for 3D modeling, computer graphics systems, and other computer-aided geometry applications.

Here is another example, this time for a polyhedron that is a rectangular block (Figure 10.15). The connectivity matrix for this polyhedron is

$$\mathbf{C}_{\text{rectangular solid}} = \begin{bmatrix} 0 & 1 & 0 & 1 & 1 & 0 & 0 & 0 \\ 1 & 0 & 1 & 0 & 0 & 1 & 0 & 0 \\ 0 & 1 & 0 & 1 & 0 & 0 & 1 & 0 \\ 1 & 0 & 1 & 0 & 0 & 0 & 0 & 1 \\ 1 & 0 & 0 & 0 & 0 & 1 & 0 & 1 \\ 0 & 1 & 0 & 0 & 1 & 0 & 1 & 0 \\ 0 & 0 & 1 & 0 & 0 & 1 & 0 & 1 \\ 0 & 0 & 0 & 1 & 1 & 0 & 1 & 0 \end{bmatrix}. \tag{10.35}$$

The connectivity matrix **C** has twice as much information as necessary. It is doubly redundant. If we draw a line diagonally from the zero in row 1 and column 1 to the zero in row 8 and column 8, then the triangular array of elements on one side of this diagonal is the mirror image of the array on the other side. It is easy to see why. For example, the entry $a_{73} = 1$ in row 7 and column 3 says that vertex 7 is joined to vertex 3 by an edge; the entry

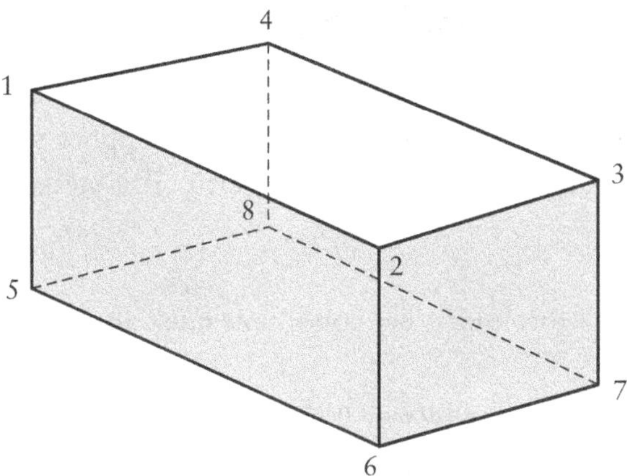

Figure 10.15 Rectangular solid connectivity

$a_{37} = 1$ in row 3 and column 7 says that vertex 3 is joined to vertex 7 with an edge. These two entries are two identical pieces of information. This redundancy is not only tolerable but also useful, because any search of these data for connectivity is more direct. It is important to note that the connectivity matrix reveals little about the actual shape of the object. With appropriate labeling of the vertices, a cube could just as easily have the same connectivity matrix as a rectangular solid or a truncated pyramid.

Data Structure for Polyhedra

Any computer program we write for geometry or computer graphics applications requires a logical method for organizing and storing the data. The following table is an example of one form of model data structure for a polyhedron. The polyhedron in this example is a cube, which, of course, has eight vertices, twelve edges, and six faces. The topology data describe how the cube's vertices, edges, and faces are connected. The circuit of vertices defines the faces in counterclockwise order as viewed from outside the cube:

Edge connectivity	Face vertex circuit
$E_1 : (\mathbf{p}_1, \mathbf{p}_2)$	$F_1 : (\mathbf{p}_1, \mathbf{p}_2, \mathbf{p}_6, \mathbf{p}_5)$
$E_2 : (\mathbf{p}_2, \mathbf{p}_3)$	$F_2 : (\mathbf{p}_3, \mathbf{p}_4, \mathbf{p}_8, \mathbf{p}_9)$
$E_3 : (\mathbf{p}_3, \mathbf{p}_4)$	$F_3 : (\mathbf{p}_5, \mathbf{p}_8, \mathbf{p}_4, \mathbf{p}_1)$
$E_4 : (\mathbf{p}_4, \mathbf{p}_1)$	$F_4 : (\mathbf{p}_2, \mathbf{p}_3, \mathbf{p}_7, \mathbf{p}_6)$
$E_5 : (\mathbf{p}_5, \mathbf{p}_6)$	$F_5 : (\mathbf{p}_5, \mathbf{p}_6, \mathbf{p}_7, \mathbf{p}_8)$
$E_6 : (\mathbf{p}_6, \mathbf{p}_7)$	$F_6 : (\mathbf{p}_4, \mathbf{p}_3, \mathbf{p}_2, \mathbf{p}_1)$
$E_7 : (\mathbf{p}_7, \mathbf{p}_8)$	
$E_8 : (\mathbf{p}_8, \mathbf{p}_5)$	
$E_9 : (\mathbf{p}_1, \mathbf{p}_5)$	
$E_{10} : (\mathbf{p}_2, \mathbf{p}_6)$	
$E_{11} : (\mathbf{p}_3, \mathbf{p}_7)$	
$E_{12} : (\mathbf{p}_4, \mathbf{p}_8)$	

Euler's Formula for Simple Polyhedra

A simple polyhedron refers to any polyhedron that can be deformed into a sphere, assuming that its faces are treated like rubber sheets. Euler's formula is both deceptively simple and powerful. It is stated as

$$V - E + F = 2. \tag{10.36}$$

To apply Euler's formula, other conditions must also be met:

- A single ring of edges bounds each face, with no holes in the faces.
- The polyhedron must have no holes through it.
- Each edge is shared by exactly two faces and is terminated by a vertex at each end.
- At least three edges must meet at each vertex.

Applying the formula to a cube yields 8 – 12 + 6 = 2 and to an octahedron yields 6 – 12 + 8 = 2. The polyhedra in Figure 10.16 satisfy the four conditions, and therefore, Euler's formula applies.

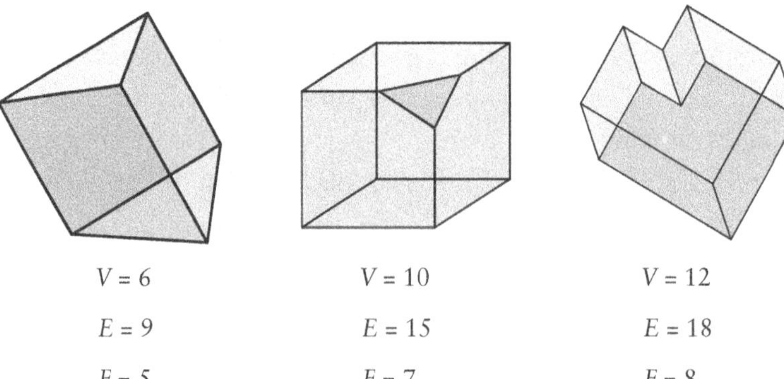

$V = 6$	$V = 10$	$V = 12$
$E = 9$	$E = 15$	$E = 18$
$F = 5$	$F = 7$	$F = 8$

Figure 10.16 Polyhedra satisfying Euler's formula

Euler's Formula for Nonsimple Polyhedra

If we subdivide a polyhedron into C polyhedral cells, then the vertices, edges, faces, and cells are related by

$$V - E + F - C = 1. \tag{10.37}$$

A cell is, itself, a closed polyhedron. Adding a vertex to the interior of a cube and joining it with edges to each of the other eight vertices creates a six-cell polyhedron with $V = 9$, $E = 20$, $F = 18$, and $C = 6$ (Figure 10.17). Note that each cell is pyramid shaped, with an external face of the cube as a base.

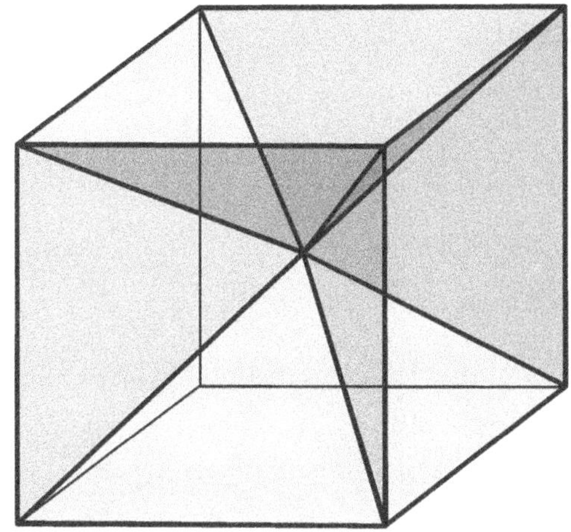

Figure 10.17 A six-cell polyhedron

If a polyhedron has one or more holes H in its faces, has passages P through it, and/or consists of disjoint bodies B, then

$$V - E + F - H + 2P = 2B. \tag{10.38}$$

A cube with a passage through it has $V = 16$, $E = 32$, $F = 16$, $H = 0$, $P = 1$, and $B = 1$ (Figure 10.18).

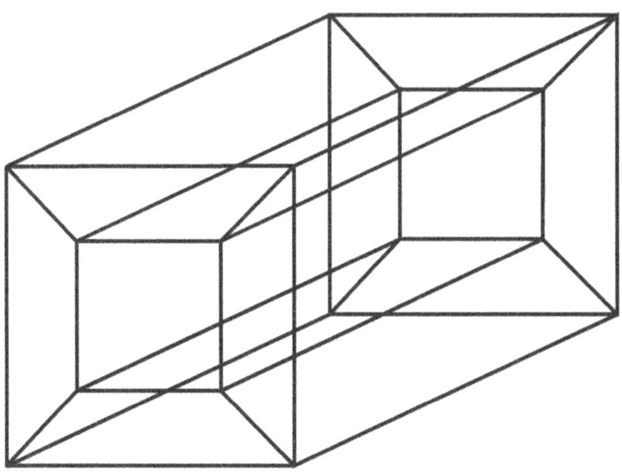

Figure 10.18 Cube with a passage through it

We can define a connectivity number n. For a sphere and all topologically equivalent shapes, $n = 0$. For torus, or doughnut-like, shapes, $n = 2$. For figure eight, or pretzel-like, shapes, $n = 4$, and so on. Therefore, we can express this as

$$V - E + F = 2 - n. \tag{10.39}$$

10.6 Exercises

10.1. Given $\mathbf{a} = [\ 6 \ \ -1 \ \ -2\ ]$, $\mathbf{b} = [\ 3 \ \ 2 \ \ 4\ ]$, and $\mathbf{c} = [\ 7 \ \ 0 \ \ 2\ ]$, write the vector equation of a line:

a. Through **a** and parallel to **b**

b. Through **b** and parallel to **c**

c. Through **c** and parallel to **a**

d. Through **a** and parallel to **a**

e. Through **b** and parallel to **a**

10.2. Find the equations of the x, y, and z vector components for the line segments given by the following pairs of endpoints:

a. $\mathbf{p}_0 = [\ 0 \ \ 0 \ \ 0\ ]$, $\mathbf{p}_1 = [\ 1 \ \ 1 \ \ 1\ ]$

b. $\mathbf{p}_0 = [\ -3 \ \ 1 \ \ 6\ ]$, $\mathbf{p}_1 = [\ 2 \ \ 0 \ \ 7\ ]$

c. $\mathbf{p}_0 = [\ 1 \ \ 1 \ \ -4\ ]$, $\mathbf{p}_1 = [\ 5 \ \ -3 \ \ 9\ ]$

d. $\mathbf{p}_0 = [\ 6 \ \ 8 \ \ 8\ ]$, $\mathbf{p}_1 = [\ -10 \ \ 0 \ \ -3\ ]$

e. $\mathbf{p}_0 = [\ 0 \ \ 0 \ \ 1\ ]$, $\mathbf{p}_1 = [\ 0 \ \ 0 \ \ -1\ ]$

10.3. Given $x = 3 + 2u$, $y = -6 + u$, and $z = 4$, find $\mathbf{p}_0$ and $\mathbf{p}_1$, with $0 \le u \le 1$.

10.4. Describe the difference between the following line segments:

For line 1, $\mathbf{p}_0 = [\ 2 \ \ 1 \ \ -2\ ]$ and $\mathbf{p}_1 = [\ 3 \ \ -3 \ \ 1\ ]$.

For line 2, $\mathbf{p}_0 = [\ 3 \ \ -3 \ \ 1\ ]$ and $\mathbf{p}_1 = [\ 2 \ \ 1 \ \ -2\ ]$.

10.5. Write the vector equation of the plane passing through **a** and parallel to **b** and **c**.

10.6. Write the vector equation of a plane that passes through the origin and is perpendicular to the y axis.

10.7. Show algebraically that the equation $\mathbf{p}_0 \bullet \mathbf{n} = d$ represents the equation of a plane passing through $\mathbf{p}_0$, normal to the unit vector **n**, where d is the perpendicular distance from the origin to the plane. Also, interpret this equation graphically with an appropriate sketch.

10.8. Show that the intersection of three planes is given by

$$\mathbf{p} = \frac{d_1(\mathbf{n}_2 \times \mathbf{n}_3) + d_2(\mathbf{n}_3 \times \mathbf{n}_1) + d_3(\mathbf{n}_1 \times \mathbf{n}_2)}{\mathbf{n}_1 \bullet \mathbf{n}_2 \times \mathbf{n}_3}.$$

10.9. Compute the distance between each of the following pairs of points:

a. (−2.7, 6.5, 0.8), (5.1, −5.7, 1.9)
b. (1, 1, 0), (4, 6, −3)
c. (7, −4, 2), (0, 2.7, −0.3)
d. (−3, 0, 0), (7, 0, 0)
e. (10, 9, −1), (3, 8, 3)

10.10. Compute the coordinates of the points in Exercise 10.9 relative to a coordinate system centered at (3, −1, 0) in the original system and parallel to it.

10.11. Compute the distance between each of the points found for Exercise 10.10.

10.12. Show that the distance between any pair of points is independent of the coordinate system chosen.

10.13. Compute the midpoint between the pairs of points given in Exercise 10.9.

10.14. Given an arbitrary set of points, find the coordinates of the vertices of a rectangular box that just encloses it.

10.15. Find the coordinates of the eight vertices of a rectangular box that just encloses the ten points given in Exercise 10.9.

10.16. Given that Δ_i is a constant for all $\mathbf{p}_i$, that is, $\Delta_i = (\Delta x_i, \Delta y_i) = (\Delta x, \Delta y)$, find $\mathbf{p}_4$ in terms of $\mathbf{p}_0$ and Δ_i.

10.17. Find the set of Δ_i's for the vertex points of a square whose sides are three units long and with $\mathbf{p}_0 = (1, 0)$. Assume that the sides of the square are parallel to the x, y coordinate axes, and proceed counterclockwise.

10.18. Repeat Exercise 10.17 for a square whose sides are four units long and with $\mathbf{p}_0 = (-2, -2)$.

10.19. Repeat Exercise 10.17 for $\mathbf{p}_0 = (1, -4)$.

10.20. Show that the line joining the midpoints of two sides of a triangle is parallel to the third side and has one-half its magnitude.

10.21. Find the midpoint of the line segment between $\mathbf{p}_0 = [\ 3 \quad 5 \quad 1\ ]$ and $\mathbf{p}_1 = [\ -2 \quad 6 \quad 4\ ]$.

10.22. Write the vector equation of a line through two points $\mathbf{p}_0$ and $\mathbf{p}_1$.

10.23. Write the vector equation of a plane containing the three noncollinear points $\mathbf{p}_0$, $\mathbf{p}_1$, and $\mathbf{p}_2$.

11 Vector-Defined Geometric Objects II

This chapter continues the discussion of vector representation of geometric objects. Here the topics are curves and surfaces, whose definitions require developing special functions. It shows how to extract local and global properties. And it sets the stage for analyzing spatial relationships (Chapter 12).

11.1 Curve-Defining Functions

We all have a strong intuitive sense of what a curve is. And although we never see a curve floating around free of any object, we can easily identify the curved edges and silhouettes of objects and just as easily imagine the curve that describes the path of a moving object. In this section we explore a way to describe a curve that relies on the mathematics of vectors. We'll look at a family of curves whose shape is determined by parametric cubic polynomial equations.

The parametric cubic equations of curves used in geometric modeling have a vector form. Now we will see how these parametric polynomial equations become vector equations, a much more intuitive and geometric way to shape and represent a curve. The following three equations define the classic polynomial form of a parametric cubic curve, producing the coordinates of points on a curve:

$$\begin{aligned} x(u) &= a_x u^3 + b_x u^2 + c_x u + d_x, \\ y(u) &= a_y u^3 + b_y u^2 + c_y u + d_y, \\ z(u) &= a_z u^3 + b_z u^2 + c_z u + d_z, \end{aligned} \tag{11.1}$$

where the parametric variable u is the independent variable assigned to a point and where $x(u)$, $y(u)$, and $z(u)$ are the cubic functions defining the dependent variables . . . the coordinates of a point on the curve.

From this set of equations, we assemble the following "algebraic" vectors:

$$\begin{aligned} \mathrm{a} &= (a_x, a_y, a_z), \\ \mathrm{b} &= (b_x, b_y, b_z), \\ \mathrm{c} &= (c_x, c_y, c_z), \\ \mathrm{d} &= (d_x, d_y, d_z). \end{aligned} \tag{11.2}$$

Put aside the temptation to think of **a**, **b**, **c**, and **d** as "geometric" vectors. That discussion comes later in this section.

Next, rewrite Equation 11.1 more compactly as the vector equation

$$\mathbf{p}(u) = \mathbf{a}u^3 + \mathbf{b}u^2 + \mathbf{c}u + \mathbf{d}. \tag{11.3}$$

The coefficients, whether in polynomial or vector form, are not the most convenient way of understanding or controlling the shape of a curve in typical modeling situations, nor do they contribute much to an intuitive sense of a curve. However, there is a way to reconstruct the form of Equation 11.3 that offers a more geometric alternative, one that allows us to define a curve segment in terms of conditions at its endpoints. This new form is the cubic Hermite curve. The boundary conditions are the endpoint coordinates and the vectors lying on the tangents at these points. The Hermite form is one of a family of curves that include Bézier and B-spline curves.

First, we will restrict the parametric variable to values in the interval from $u = 0$ to $u = 1$, producing, in effect, a point-bounded curve segment. Each u-value in this interval generates a point on the curve. So using the notation of Equation 11.3, with the endpoints $\mathbf{p}(0)$ and $\mathbf{p}(1)$ and the corresponding tangent vectors $\mathbf{p}^u(0)$ and $\mathbf{p}^u(1)$, we obtain the following four equations:

$$\begin{aligned} \mathbf{p}(0) &= \mathbf{d}, \\ \mathbf{p}(1) &= \mathbf{a} + \mathbf{b} + \mathbf{c} + \mathbf{d}, \\ \mathbf{p}^u(0) &= \mathbf{c}, \\ \mathbf{p}^u(1) &= 3\mathbf{a} + 2\mathbf{b} + \mathbf{c}, \end{aligned} \tag{11.4}$$

where substituting $u = 0$ into Equation 11.3 produces $\mathbf{p}(0)$ and substituting $u = 1$ into this equation obtains $\mathbf{p}(1)$. [Remember, the notation $\mathbf{p}^u$ represents the first derivative of $\mathbf{p}(u)$ with respect to the independent variable u, producing the tangent vector at u.]

Finally, differentiating $\mathbf{p}(u)$ with respect to u yields $\mathbf{p}^u(u) = 3\mathbf{a}u^2 + 2\mathbf{b}u + \mathbf{c}$, and substituting $u = 0$ and $u = 1$ into this obtains $\mathbf{p}^u(0)$ and $\mathbf{p}^u(1)$. Solving this set of four simultaneous equations in four unknowns yields the polynomial coefficients **a**, **b**, **c**, and **d** in terms of the boundary condition vectors:

$$\begin{aligned} \mathbf{a} &= 2\mathbf{p}(0) - 2\mathbf{p}(1) + \mathbf{p}^u(0) + \mathbf{p}^u(1), \\ \mathbf{b} &= -3\mathbf{p}(0) + 3\mathbf{p}(1) - 2\mathbf{p}^u(0) - \mathbf{p}^u(1), \\ \mathbf{c} &= \mathbf{p}^u(0), \\ \mathbf{d} &= \mathbf{p}(0). \end{aligned} \tag{11.5}$$

Substituting these expressions for the polynomial coefficients into Equation 11.3 and rearranging terms produces

$$\begin{aligned}\mathbf{p}(u) &= \left(2u^3-3u^2+1\right)\mathbf{p}(0)+\left(-2u^3+3u^2\right)\mathbf{p}(1)\\ &\quad+\left(u^3-2u^2+u\right)\mathbf{p}^u(0)+\left(u^3-u^2\right)\mathbf{p}^u(1),\end{aligned} \tag{11.6}$$

which means that given two endpoints and their corresponding tangent vectors, this vector equation defines points along the curve. More advanced textbooks on geometric modeling reveal other features of the Hermite curve. And it turns out that under most conditions it is possible to convert between it and Bézier and B-spline curves.

There is more to these boundary tangent vectors than is immediately apparent, because in another equivalent form they reveal more degrees of freedom to control curve shape. Here is that form:

$$\begin{aligned}&\mathbf{p}^u(0) = k_0\mathbf{t}(0) \text{ for } u = 0,\\ &\mathbf{p}^u(1) = k_1\mathbf{t}(1) \text{ for } u = 1,\end{aligned} \tag{11.7}$$

where $\mathbf{t}(u)$ is a unit vector given by

$$\mathbf{t}(u) = \mathbf{p}^u(u)/|\mathbf{p}^u(u)|. \tag{11.8}$$

This means that we can keep the tangent vector directions constant, using the unit tangent vectors, and have the freedom to change their magnitudes and so change the shape of the curve.

How did that happen? Here is some more detail: The direction cosines of the tangent to a point u on a curve comprise the components of the unit vector $\mathbf{t}(u)$, where the direction cosines are $t_x(u)$, $t_y(u)$, and $t_z(u)$. From elementary geometry, we know that

$$\begin{aligned}|\mathbf{t}(u)| &= \sqrt{t_x^2(u)+t_y^2(u)+t_z^2(u)}\\ &= 1,\end{aligned} \tag{11.9a}$$

where $t_x(u)$, $t_y(u)$, and $t_z(u) \in [0, 1]$.

Furthermore, because $k = |\mathbf{p}^u(u)|$, then

$$\mathbf{p}^u(u) = k\mathbf{t}(u), \tag{11.9b}$$

where k represents the magnitude of $\mathbf{p}^u(u)$.

The subscripts indicate a specific point, corresponding to its u-value, so that, as we have seen, the tangent vectors for the two endpoints of a curve are $\mathbf{p}_0^u = k_0\mathbf{t}_0$ and $\mathbf{p}_1^u = k_1\mathbf{t}_1$. Because the three direction cosines are related by $|\mathbf{t}(u)| \equiv 1$, only two are independent variables. So the shape of a (Hermite) parametric cubic curve springs from these boundary conditions (Figure 11.1):

- The endpoints $\mathbf{p}(0)$ and $\mathbf{p}(1)$
- The unit tangent vectors at the endpoints, $\mathbf{t}(0)$ and $\mathbf{t}(1)$
- The tangent vector magnitudes k_0 and k_1

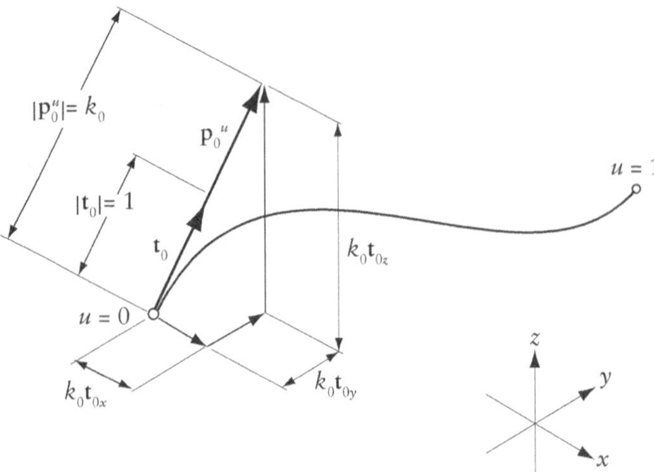

Figure 11.1 Tangent vector elements

This is a very interesting state of affairs, because it shows that it is a property of cubic Hermite curves that many different curves are possible (in fact, an infinite family of them), all of which have the same endpoints and slopes and yet entirely different interior shapes, depending on the tangent vector magnitudes k_0 and k_1. By varying k_0 and k_1, we can control the curvature at each end, fix the location of some intermediate point, or satisfy some other shape criteria (Figure 11.2).

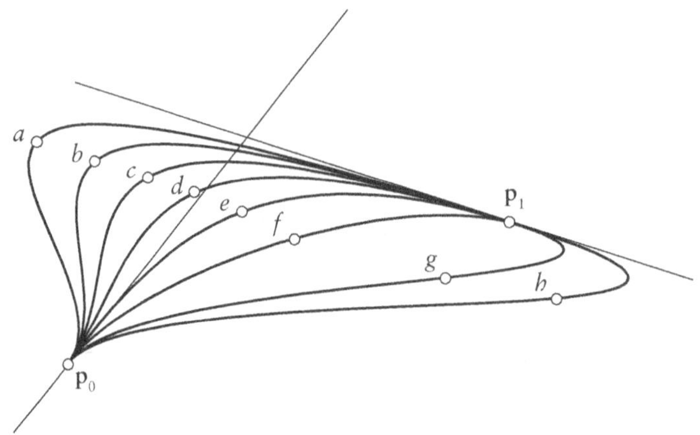

Figure 11.2 Effect of tangent vector magnitudes on the shape of the curve

11.2 Bézier Curves

A set of control points given as vectors defines the shape of a Bézier curve. The number of control points determines the degree of the parametric functions that underlie this curve: Three points define a second-degree curve, four points define a cubic curve, and so forth. A cubic Bézier curve begins on the first control point and ends on the last. The two intermediate control points determine the tangent vectors at each end and the interior shape of the curve.

Let's look at a simple example (Figure 11.3). The vectors, or points, if you prefer, $\mathbf{p}_0$, $\mathbf{p}_1$, $\mathbf{p}_2$, and $\mathbf{p}_3$ define the control points and also the convex hull that encloses the curve. The order of the points indicates the direction in which the parametric variable is increasing. (The cubic parametric functions are similar to those of the Hermite cubic functions for curves. And, indeed, the two types of curves are interchangeable. There is a simple way to transform one into the other. This is covered in more advanced textbooks.)

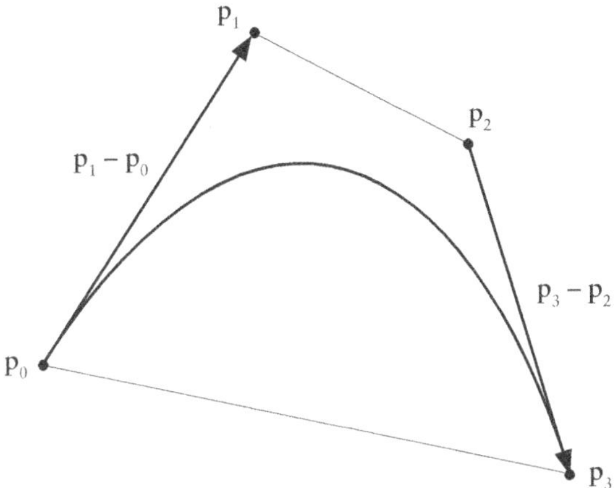

Figure 11.3 A simple Bézier curve

The tangent vector at $\mathbf{p}_0$ is $\mathbf{p}_1 - \mathbf{p}_0$. At $\mathbf{p}_3$ the tangent vector is $\mathbf{p}_3 - \mathbf{p}_2$. We can change the interior shape of a Bézier curve by moving the points $\mathbf{p}_1$ and $\mathbf{p}_2$. For example, we can preserve the tangent line at $\mathbf{p}_0$ by moving $\mathbf{p}_1$ along the tangent line (Figure 11.4).

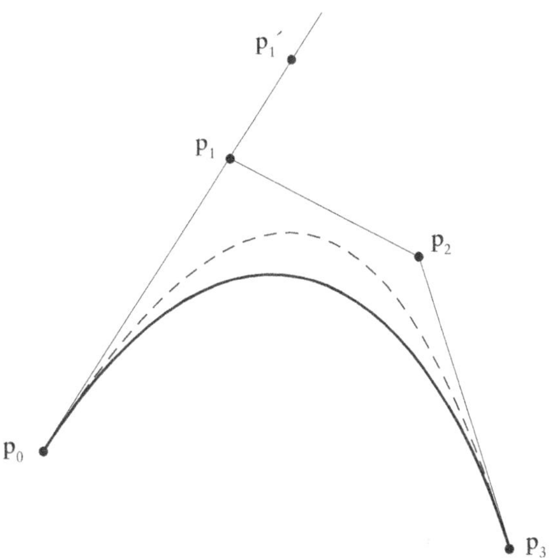

Figure 11.4 Changing the shape of a Bézier curve

Here is what the vector equation of a Bézier curve looks like (its derivation can be found in more advanced textbooks):

$$\mathbf{p}(u) = (1-u)^3\mathbf{p}_0 + 3u(1-u)^2\mathbf{p}_1 + 3u^2(1-u)\mathbf{p}_2 + u^3\mathbf{p}_3.$$

11.3 Local Properties of a Curve

Now let's look at some properties of curves. In general, what follows is independent of any specific breed of curve. It applies to Hermite, Bézier, b-spline, NURBS, and so forth, as long as the functions defining points on the curve are differentiable.

There are three characteristic vectors and three vector-defined planes associated with each point on a curve that play an important role in geometric modeling. They are local properties because they vary from point to point on a curve. Figure 11.5 assembles these vectors and planes and illustrates that these elements form a local, three-dimensional orthogonal coordinate system consisting of three axis vectors and three coordinate planes. This local coordinate system is the moving trihedron, or Frénet frame, of the curve, which is fundamental to the study of the differential geometry of curves and is an important ingredient in some sweep representations of solids in geometric and 3D modeling. We will explore these and other local properties of a curve, most often expressed in terms of vectors, matrices, and determinants at a point $\mathbf{p}_i$ or $\mathbf{p}(u_i)$ on a curve.

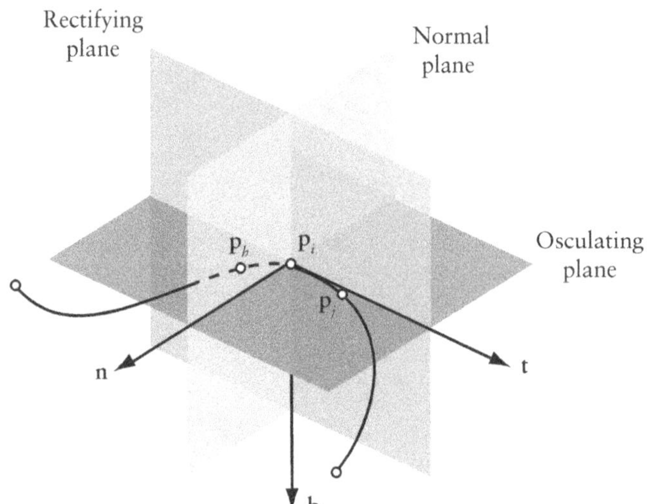

Figure 11.5 The moving trihedron

As usual, denote the tangent vector at point $\mathbf{p}_i$ on a curve by $\mathbf{p}_i^u$. It is necessary in many situations to work with a unit tangent vector at $\mathbf{p}_i$, where

$$\mathbf{t}_i = \mathbf{p}_i^u / \left|\mathbf{p}_i^u\right|, \tag{11.10}$$

or sometimes denoted as $\hat{\mathbf{p}}_i^u$.

The equation of a straight line through $\mathbf{p}_i$ and parallel to $\mathbf{t}_i$ is

$$\mathbf{q} = \mathbf{p}_i + a\mathbf{t}_i, \tag{11.11}$$

where a is a scalar determining the distance along $\mathbf{t}_i$ from $\mathbf{p}_i$ (Figure 11.6).

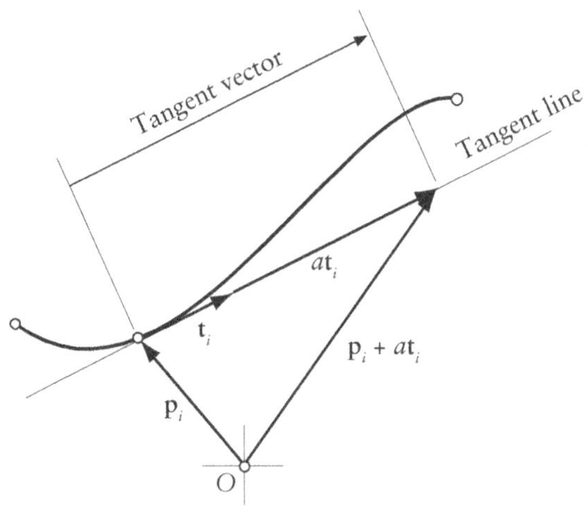

Figure 11.6 Tangent vector and line

The normal plane at a point $\mathbf{p}_i$ on a curve is the plane through $\mathbf{p}_i$ perpendicular to the vector $\mathbf{t}_i$. The equation of the normal plane is the scalar product

$$(\mathbf{q} - \mathbf{p}_i) \bullet \mathbf{t}_i = 0, \tag{11.12}$$

where $\mathbf{q}$ is any point on the plane. Since the vector $\mathbf{q} - \mathbf{p}_i$ lies in the plane and $\mathbf{q}$ is any point on the plane, we assert that if $\mathbf{q} - \mathbf{p}_i$ is perpendicular to $\mathbf{t}_i$, then the plane is also perpendicular to $\mathbf{t}_i$ (Figure 11.7).

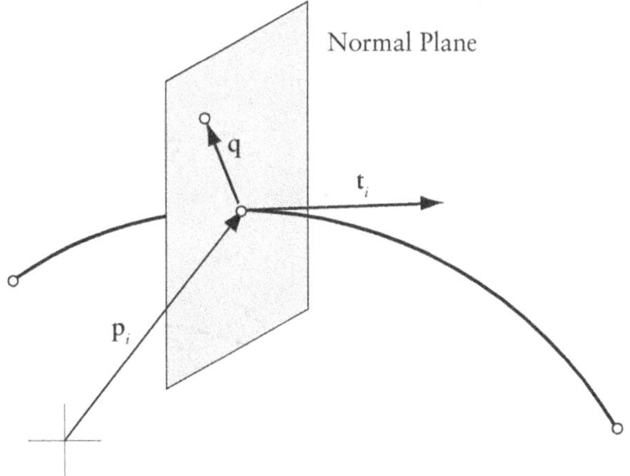

Figure 11.7 Normal plane

We can also write Equation 11.12 as

$$(\mathbf{q} - \mathbf{p}_i) \bullet \mathbf{p}_i^u = 0, \tag{11.13}$$

because $\mathbf{t}_i$ and $\mathbf{p}_i^u$ are parallel. Denoting the coordinates of point $\mathbf{q}$ as x, y, z, we rewrite Equation 11.13 as

$$\begin{bmatrix} x-x_i \\ y-y_i \\ z-z_i \end{bmatrix}^{\mathrm{T}} \begin{bmatrix} x_i^u \\ y_i^u \\ z_i^u \end{bmatrix} = 0, \tag{11.14}$$

or

$$(x-x_i)x_i^u + (y-y_i)y_i^u + (z-z_i)z_i^u = 0. \tag{11.15}$$

Rearranging the terms into a form analogous to the familiar equation for a plane, $Ax + By + Cz + D = 0$, produces

$$xx_i^u + yy_i^u + zz_i^u - (x_i x_i^u + y_i y_i^u + z_i z_i^u) = 0. \tag{11.16}$$

The principal normal vector at a point $\mathbf{p}_i$ on a curve must lie in the normal plane at that point. However, it is a special normal vector, among the many possible, in that it points in the direction in which the curve is turning and toward the center of curvature at that point. The problem is to find this vector given the parametric expression for a curve.

The construction in Figure 11.8a demonstrates that in moving an infinitesimal distance from point to point along the curve in the neighborhood of $\mathbf{p}_i$, the tangent vector $\mathbf{p}_i^u$ swings in a direction given by $\mathbf{p}^{uu}$. This action suggests the merest hint of a plane, the osculating plane. The principal normal vector lies on the intersection of this plane and the normal plane.

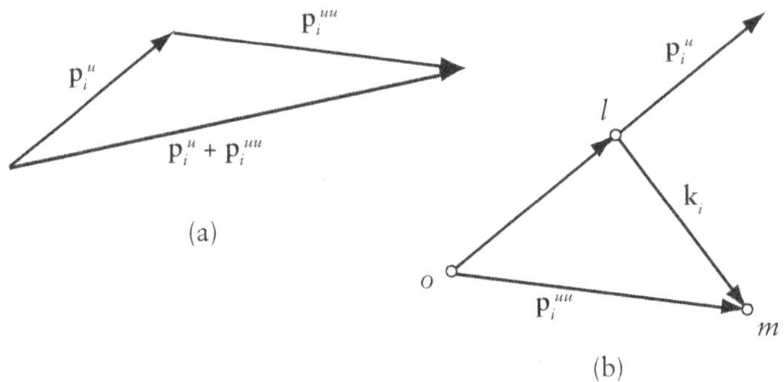

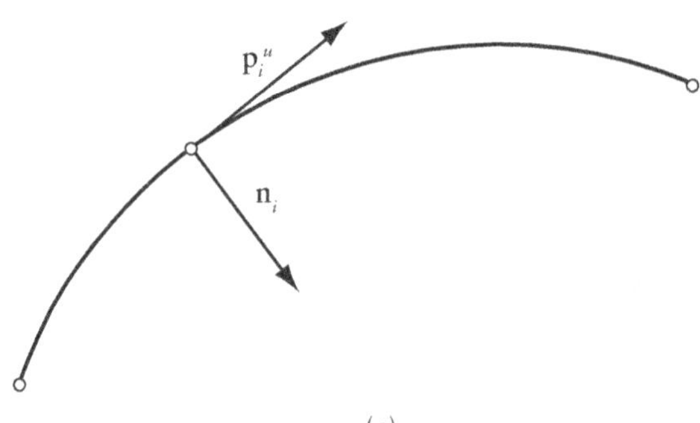

Figure 11.8 Principal normal vector and line

Now we must find a vector normal to $\mathbf{p}^u$ and in the plane of $\mathbf{p}_i^u$ and $\mathbf{p}_i^{uu}$, proceeding as follows: From elementary vector geometry, the projection of $\mathbf{p}_i^{uu}$ onto $\mathbf{p}_i^u$ is the scalar product:

$$\text{Length of the projection of } \mathbf{p}_i^{uu} \text{ onto } \mathbf{p}_i^u = \mathbf{p}_i^{uu} \bullet \left(\frac{\mathbf{p}_i^u}{\left|\mathbf{p}_i^u\right|} \right), \tag{11.17}$$

and the line of projection lm is normal to $\mathbf{p}_i^u$ (Figure 11.8b). Using this projected length, a scalar, we construct the vector along $0l$ by multiplying the unit vector $\mathbf{p}_i^u / | \mathbf{p}_i^u |$ by the length found in Equation 11.17, so that

$$\overline{Ol} = \frac{\mathbf{p}_i^{uu} \bullet \mathbf{p}_i^u}{\left|\mathbf{p}_i^u\right|^2} \mathbf{p}_i^u . \tag{11.18}$$

Next, letting $\mathbf{k}_i$ denote the vector from l to m and applying simple vector algebra, we find that

$$\mathbf{k}_i = \mathbf{p}_i^{uu} - \frac{\mathbf{p}_i^{uu} \bullet \mathbf{p}_i^u}{\left|\mathbf{p}_i^u\right|^2} \mathbf{p}_i^u . \tag{11.19}$$

This is most of the solution; the rest is shown in Figure 11.8c, where the principal unit normal vector is

$$\mathbf{n}_i = \frac{\mathbf{k}_i}{\left|\mathbf{k}_i\right|} . \tag{11.20}$$

There is an important well-known theorem from vector calculus that states that if $\mathbf{t}$ is a unit vector function, then $d\mathbf{t}/du$ is orthogonal to $\mathbf{t}$. This direct computation yields $\mathbf{n}$. The problem with this approach is in establishing a readily differentiable function $\mathbf{t}(u)$, because, in this case, it is defined by $\mathbf{p}^u / | \mathbf{p}^u |$, which must be differentiated before evaluating the expression at u_i. Establishing $\mathbf{t}(u)$ is usually difficult and often impossible, depending on the nature of the parametric function defining the curve.

For a plane curve, we apply Equations 11.19 and 11.20 to find the principal normal at a point. There is another approach, although it has limitations; it is analogous to the method for finding the normal plane. For simplicity, we assume the curve lies in the xy plane, and then the following must be true:

$$\mathbf{n}_i \bullet \mathbf{t}_i = 0, \tag{11.21}$$

or, of course,

$$\mathbf{n}_i \bullet \mathbf{p}_i^u = 0. \tag{11.22}$$

Writing Equation 11.21 in expanded component form yields

$$n_x t_x + n_y t_y = 0. \tag{11.23}$$

Treating $\mathbf{n}_i$ as a unit vector, we have

$$n_x^2 + n_y^2 = 1. \tag{11.24}$$

Because we assume to know t_x and t_y, we readily solve for n_x and n_y. However, in doing this, we find that an ambiguity arises in the signs of n_x and n_y. We have found the line of action of $\mathbf{n}$ but cannot determine in which direction along this line the vector is pointing.

Finally, we compute the principal normal line at a point $\mathbf{p}_i$ on a curve

$$\mathbf{q} = \mathbf{p}_i + a\mathbf{n}_i, \tag{11.25}$$

where $\mathbf{q}$ is any point on the line and a is a scalar determining the distance and direction along $\mathbf{n}_i$ from $\mathbf{p}_i$.

Another vector normal to the curve at a point $\mathbf{p}_i$ and lying in the normal plane is the binormal vector. We make direct use of the principal normal and tangent vectors to define the binormal vector $\mathbf{b}_i$ by simply computing their vector product:

$$\mathbf{b}_i = \mathbf{t}_i \times \mathbf{n}_i. \tag{11.26}$$

This formulation ensures that $\mathbf{b}_i$ is normal to $\mathbf{t}_i$ and that it lies in the normal plane. It is also normal to $\mathbf{n}_i$, so the three vectors form an orthogonal frame with considerable significance for geometric modeling.

The osculating plane at a point $\mathbf{p}_i$ on a curve is the limiting position of the plane defined by $\mathbf{p}_i$ and two neighboring points $\mathbf{p}_h$ and $\mathbf{p}_j$ on the curve, as these neighbor points independently approach $\mathbf{p}_i$ (Figure 11.9). Notice that the three points cannot be collinear and that the tangent vector $\mathbf{p}_i^u$ lies in this plane. The equation of the osculating plane is given by the following determinant:

$$\begin{vmatrix} (\mathbf{q} - \mathbf{p}_i) & \mathbf{p}_i^u & \mathbf{p}_i^{uu} \end{vmatrix} = 0, \tag{11.27}$$

where $\mathbf{q}$, again, is any generic point on the osculating plane. Writing the vectors in terms of their components produces

$$\begin{vmatrix} x - x_i & x_i^u & x_i^{uu} \\ y - y_i & y_i^u & y_i^{uu} \\ z - z_i & z_i^u & z_i^{uu} \end{vmatrix} = 0. \tag{11.28}$$

Expanding this determinant yields

$$\begin{aligned} &(x - x_i)\left(y_i^u z_i^{uu} - y_i^{uu} z_i^u\right) \\ &-(y - y_i)\left(x_i^u z_i^{uu} - x_i^{uu} z_i^u\right) \\ &+(z - z_i)\left(x_i^u y_i^{uu} - x_i^{uu} y_i^u\right) = 0. \end{aligned} \tag{11.29}$$

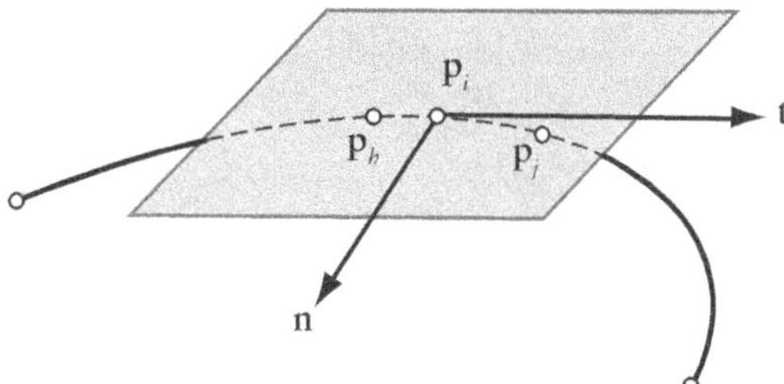

Figure 11.9 Osculating plane

Another way of formulating the osculating plane is by using the vectors **t** and **n**, which, from other considerations, we know must lie in the plane. Therefore,

$$\mathbf{q} = \mathbf{p}_i + u\mathbf{t}_i + w\mathbf{n}_i, \tag{11.30}$$

where **q** is any point on the osculating plane and u and w are parametric variables of unspecified bounds.

The rectifying plane at a point $\mathbf{p}_i$ is the plane through $\mathbf{p}_i$ and perpendicular to the principal normal $\mathbf{n}_i$, so that

$$(\mathbf{r} - \mathbf{p}_i) \bullet \mathbf{n}_i = 0, \tag{11.31}$$

where **r** is a generic point on the rectifying plane.

Another way of formulating this plane is analogous to the method we just developed for the osculating plane. It is

$$\mathbf{r} = \mathbf{p}_i + u\mathbf{t}_i + w\mathbf{b}_i, \tag{11.32}$$

where **r** is, again, any point on the rectifying plane and u and w are parametric variables of unspecified bounds.

The curvature $1/\rho_i$ at a point $\mathbf{p}_i$ on a curve is

$$\frac{1}{\rho_i} = \frac{\left|\mathbf{p}_i^u \times \mathbf{p}_i^{uu}\right|}{\left|\mathbf{p}_i^u\right|^3}, \tag{11.33}$$

where ρ_i is the radius of curvature. We also use κ to denote curvature, where $\kappa = 1/\rho$ and curvature is measured in the osculating plane along the principal normal vector $\mathbf{n}_i$. We readily construct the curvature vector $\mathbf{k} = \rho_i\mathbf{n}_i$, which is the vector from $\mathbf{p}_i$ to the center of curvature (Figure 11.10).

The curvature of a plane curve in the xy plane is

$$\frac{1}{\rho} = \frac{d^2y/dx^2}{\left[1+(dy/dx)^2\right]^{3/2}}. \tag{11.34}$$

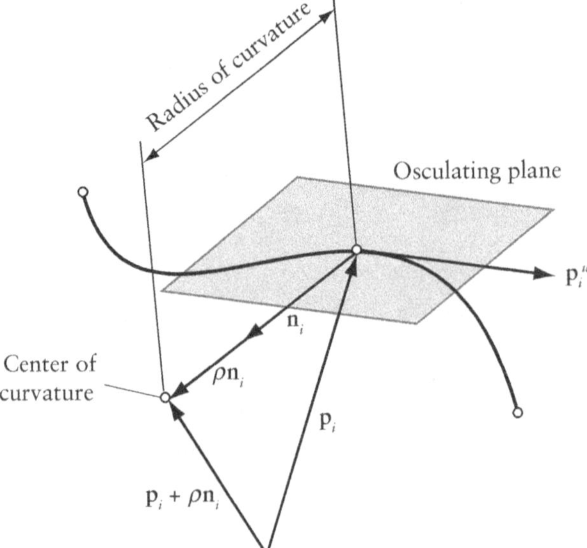

Figure 11.10 Curvature

We may now write Equation 11.34 in terms of the parametric derivatives, where

$$\frac{dy}{dx} = \frac{y^u}{x^u} \tag{11.35}$$

and

$$\begin{aligned} \frac{d^2y}{dx^2} &= \frac{d}{du}\left(\frac{y^u}{x^u}\right)\frac{du}{dx} \\ &= \frac{x^u y^{uu} - y^u x^{uu}}{\left(x^u\right)^3}, \end{aligned} \tag{11.36}$$

so that

$$\begin{aligned} \left[1+\left(\frac{dy}{dx}\right)^2\right]^{3/2} &= \left[1+\left(\frac{y^u}{x^u}\right)^2\right]^{3/2} \\ &= \frac{1}{\left(x^u\right)^3}\left[\left(x^u\right)^2+\left(y^u\right)^2\right]^{3/2}. \end{aligned} \tag{11.37}$$

We substitute the preceding results into Equation 11.34 to obtain

$$\frac{1}{\rho} = \frac{x^u y^{uu} - x^{uu} y^u}{\left[\left(x^u\right)^2+\left(y^u\right)^2\right]^{3/2}}, \tag{11.38}$$

which is the two-dimensional scalar equivalent of Equation 11.33.

The torsion at a point $\mathbf{p}_i$ on a curve is the limit of the ratio of the angle between the binormal at $\mathbf{p}_i$ and the binormal at a neighboring point $\mathbf{p}_h$ to the arc length of the curve between $\mathbf{p}_h$ and $\mathbf{p}_i$, as $\mathbf{p}_h$ approaches $\mathbf{p}_i$ along the curve

(Figure 11.11). We can see this clearly by observing the rectifying planes. Torsion is a rotation or twist about the tangent vector, given by the formula

$$\tau_i = \frac{\left[\begin{array}{ccc} \mathbf{p}_i^u & \mathbf{p}_i^{uu} & \mathbf{p}_i^{uuu} \end{array}\right]}{\left|\mathbf{p}_i^u \times \mathbf{p}_i^{uu}\right|^2}, \tag{11.39}$$

where the numerator is a scalar triple product of the three vectors.

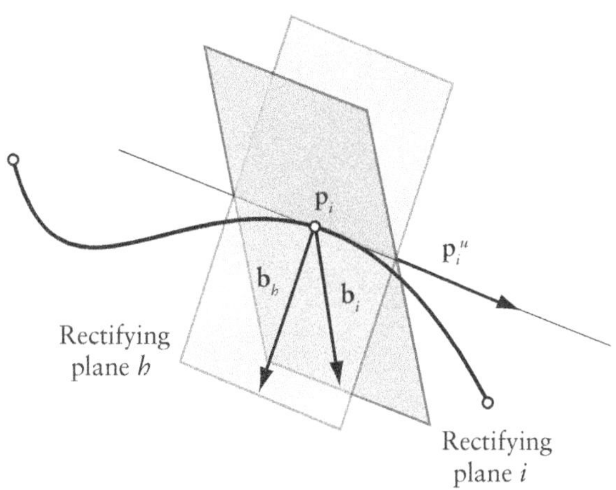

Figure 11.11 Torsion

It is not surprising to find that curvature and torsion are related. From Equation 11.33, we have

$$\left|\mathbf{p}_i^u \times \mathbf{p}_i^{uu}\right| = \frac{\left|\mathbf{p}_i^u\right|^3}{\rho_i}, \tag{11.40}$$

so that

$$\left|\mathbf{p}_i^u \times \mathbf{p}_i^{uu}\right|^2 = \frac{\left|\mathbf{p}^i\right|^6}{\rho_i^2}. \tag{11.41}$$

Substituting this equation into Equation 11.39, we obtain

$$\tau_i = \frac{\left[\begin{array}{ccc} \mathbf{p}_i^u & \mathbf{p}_i^{uu} & \mathbf{p}_i^{uuu} \end{array}\right]}{(1/\rho_i)^2\left|\mathbf{p}_i^u\right|^6}. \tag{11.42}$$

We can clean up Equation 11.42 somewhat as follows:

$$\tau_i = \frac{\left[\begin{array}{ccc} \mathbf{p}_i^u & \mathbf{p}_i^{uu} & \mathbf{p}_i^{uuu} \end{array}\right]}{\left|\mathbf{p}_i^u\right|^6}\rho_i^2. \tag{11.43}$$

This relationship holds only if $1/\rho \neq 0$. Furthermore, if the curve is a plane curve, then $\tau = 0$.

Points on a curve where the curvature equals zero are inflection points. Curvature is zero when the numerator in Equation 11.33 is zero. Thus, we find the inflection points of a curve, if any, by solving

$$\left|\mathbf{p}_i^u \times \mathbf{p}_i^{uu}\right| = 0. \tag{11.44}$$

From elementary vector geometry, we know that if two vectors are parallel, then their vector product is zero. Therefore, any point on a curve with $\mathbf{p}_i^u$ parallel to $\mathbf{p}_i^{uu}$ is an inflection point. Let us see what Equation 11.44 can show us. First, we drop the subscript i, because we are not dealing with a specific point. Then we rewrite the equation using the well-known vector relationship $|\mathbf{a}| = \sqrt{\mathbf{a} \bullet \mathbf{a}}$, producing

$$\begin{aligned} \left|\mathbf{p}^u \times \mathbf{p}^{uu}\right| &= \sqrt{\left(\mathbf{p}^u \times \mathbf{p}^{uu}\right) \bullet \left(\mathbf{p}^u \times \mathbf{p}^{uu}\right)} \\ &= 0, \end{aligned} \tag{11.45}$$

or more simply,

$$\left(\mathbf{p}^u \times \mathbf{p}^{uu}\right) \bullet \left(\mathbf{p}^u \times \mathbf{p}^{uu}\right) = 0. \tag{11.46}$$

Next, we apply the scalar product operation to obtain

$$\left(\mathbf{p}^u \times \mathbf{p}^{uu}\right)_x^2 + \left(\mathbf{p}^u \times \mathbf{p}^{uu}\right)_y^2 + \left(\mathbf{p}^u \times \mathbf{p}^{uu}\right)_z^2 = 0, \tag{11.47}$$

where $(\mathbf{p}^u \times \mathbf{p}^{uu})_x$ denotes the x component of the vector resulting from the vector product operation indicated within parentheses; similarly for the y, z components.

We perform the indicated vector product and obtain

$$\begin{aligned} &\left(y^u z^{uu} - z^u y^{uu}\right)^2 \\ &+\left(z^u x^{uu} - x^u z^{uu}\right)^2 \\ &+\left(x^u y^{uu} - y^u x^{uu}\right)^2 = 0. \end{aligned} \tag{11.48}$$

Equation 11.48 tells us that the sum of the numerators of parametric expressions for d^2y/dx^2, d^2x/dz^2, d^2z/dy^2 must equal zero. There is an even deeper meaning here. Equation 11.48 is also telling us that $d^2y/dx^2 = 0$, $d^2x/dz^2 = 0$, $d^2z/dy^2 = 0$, and each second-derivative expression is identically equal to zero. To satisfy this condition, only the numerators in Equation 11.36 and the two other similar expressions must be zero. In fact, each of the three terms in Equation 11.48 must be zero. Thus,

$$y^u z^{uu} - z^u y^{uu} = 0, \tag{11.49}$$

$$z^u x^{uu} - x^u z^{uu} = 0, \tag{11.50}$$

$$x^u y^{uu} - y^u x^{uu} = 0. \tag{11.51}$$

For parametric cubic curves, the expressions reduce to quadratic polynomials. For example, expanded in terms of the cubic Hermite basis, the last of these equations becomes

$$3(b_x a_y - a_x b_y)u^2 + 3(c_x a_y - a_x c_y)u + (c_x b_y - b_x c_y) = 0. \tag{11.52}$$

Thus, we must find the roots of three quadratic equations in u in the interval $u \in [0, 1]$. If any roots exist in this interval, we must compare them. Only roots common to all three determine an inflection point. Two roots are possible, and therefore, two distinct inflection points are possible.

11.4 Global Properties of a Curve

The global properties of a curve include its length and whether or not it is a plane curve or a straight line. Other properties include the presence or absence of inflection points and the existence and location of extreme or min/max points (relative to some coordinate system, of course).

The simplest approach to computing the arc length of a parametric curve between parameter values u_1 and u_2 is to first divide this segment into n equal parametric intervals. We compute the x, y, z coordinates at the endpoint of each interval and then simply compute the sum of the successive straight-line distances between these points. We express this as

$$L = \sum_{i=1}^{n} l_i, \tag{11.53}$$

where

$$l_i = \sqrt{(\mathbf{p}_i - \mathbf{p}_{i-1}) \bullet (\mathbf{p}_i - \mathbf{p}_{i-1})}. \tag{11.54}$$

This is not usually computationally efficient, nor is it particularly accurate unless n is very large, because this procedure always returns a somewhat shorter-than-true arc length. Therefore, let us consider another method.

The arc length between $\mathbf{p}(u_1)$ and $\mathbf{p}(u_2)$ is

$$L = \int_{u_1}^{u_2} \sqrt{\mathbf{p}^u \bullet \mathbf{p}^u}\, du, \tag{11.55}$$

where $u_2 > u_1$.

When we complete the vector operation, this equation becomes

$$L = \int_{u_1}^{u_2} \sqrt{a_4 u^4 + a_3 u^3 + a_2 u^2 + a_1 u + a_0}\, du. \tag{11.56}$$

We define the a_i constants in terms of the algebraic coefficients.

Using the more general function notation for Equation 11.56, we obtain

$$L = \int_{u_1}^{u_2} f(u)\,du. \tag{11.57}$$

Gaussian quadrature, a numerical analysis tool, yields

$$\int_{u_1}^{u_2} f(u)\,du = \sum_{i=1}^{n} w_i f(u_i), \tag{11.58}$$

where n is the number of points, w_i are the weight values, and u_i are the Gaussian abscissas (not shown here).

We can normalize the abscissas to a more convenient interval, $u_i \in [0, 1]$, by using the following transformation:

$$t = \frac{u - u_i}{u_2 - u_1}. \tag{11.59}$$

Now we can express L as

$$L = (u_2 - u_1)\int_0^1 f\left[u_1 + (u_2 - u_1)t\right]dt, \tag{11.60}$$

or

$$L = (u_2 - u_1)\sum_{i=1}^{n} w_i g(t_i), \tag{11.61}$$

where the weights and abscissas are taken with respect to the new interval.

We can apply two simple tests to a curve to determine if it possesses special characteristics. The test to determine if a curve is a plane curve uses the observation that a plane curve has zero torsion or twist, $\tau = 0$. This means that the numerator in Equation 11.39 must equal zero, so that

$$\begin{vmatrix} \mathbf{p}^u & \mathbf{p}^{uu} & \mathbf{p}^{uuu} \end{vmatrix} = 0, \tag{11.62}$$

which, of course, expands to

$$\begin{vmatrix} x^u & x^{uu} & x^{uuu} \\ y^u & y^{uu} & y^{uuu} \\ z^u & z^{uu} & z^{uuu} \end{vmatrix} = 0. \tag{11.63}$$

Finding that a curve is planar, we next want to know if it is a straight line. In order for this to be true, the curvature must be zero at all points on the curve, which means that

$$\left|\mathbf{p}^u \times \mathbf{p}^{uu}\right| = 0. \tag{11.64}$$

There are methods to analyze planar parametric cubic curves to determine the conditions producing loops, cusps, or inflection points. The presence of a loop means that the curve is self-intersecting, a cusp is a point on a curve where the tangent vector is discontinuous, and an inflection point is a point of zero curvature on a curve.

11.5 Surface-Defining Functions

The first section shows how parametric cubic polynomials and their vector equivalents define curve segments. Now we'll see how, in a similar way, parametric bicubic polynomials define surface patches. And these polynomials also have vector equivalents. Here is how this happens.

The simplest and most common surface element in geometric modeling is a patch. (Note that the term "patch" suggests that several patches may be joined along their bounding edges to form a larger, more complex surface than that possible using a single patch.) A patch is a four-sided piece of a surface whose boundary and internal points are given by bicubic (usually) polynomial functions of the general form

$$\begin{aligned} x(u,w) = {} & a_{33x}u^3w^3 + a_{32x}u^3w^2 + a_{31x}u^3w + a_{30x}u^3 \\ & + a_{23x}u^2w^3 + a_{22x}u^2w^2 + a_{21x}u^2w + a_{20x}u^2 \\ & + a_{13x}uw^3 + a_{12x}uw^2 + a_{11x}uw + a_{10x}u \\ & + a_{03x}w^3 + a_{02x}w^2 + a_{01x}w + a_{00x}, \end{aligned} \tag{11.65}$$

$$\begin{aligned} y(u,w) = {} & a_{33y}u^3w^3 + a_{32y}u^3w^2 + a_{31y}u^3w + a_{30y}u^3 \\ & + a_{23y}u^2w^3 + a_{22y}u^2w^2 + a_{21y}u^2w + a_{20y}u^2 \\ & + a_{13y}uw^3 + a_{12y}uw^2 + a_{11y}uw + a_{10y}u \\ & + a_{03y}w^3 + a_{02y}w^2 + a_{01y}w + a_{00y}, \end{aligned} \tag{11.66}$$

$$\begin{aligned} z(u,w) = {} & a_{33z}u^3w^3 + a_{32z}u^3w^2 + a_{31z}u^3w + a_{30z}u^3 \\ & + a_{23z}u^2w^3 + a_{22z}u^2w^2 + a_{21z}u^2w + a_{20z}u^2 \\ & + a_{13z}uw^3 + a_{12z}uw^2 + a_{11z}uw + a_{10z}u \\ & + a_{03z}w^3 + a_{02z}w^2 + a_{01z}w + a_{00z}, \end{aligned} \tag{11.67}$$

where $u, w \in [0, 1]$.

This set of parametric equations generates a four-sided patch, with four corner points, four boundary curves, and assorted tangent vectors. What happens at these boundaries determines the interior shape of the patch. There are 48 coefficients in these equations whose values determine the shape of the patch. As with curves, they have, after some algebraic manipulations not done here, vector equivalents. Figure 11.12 shows some of these.

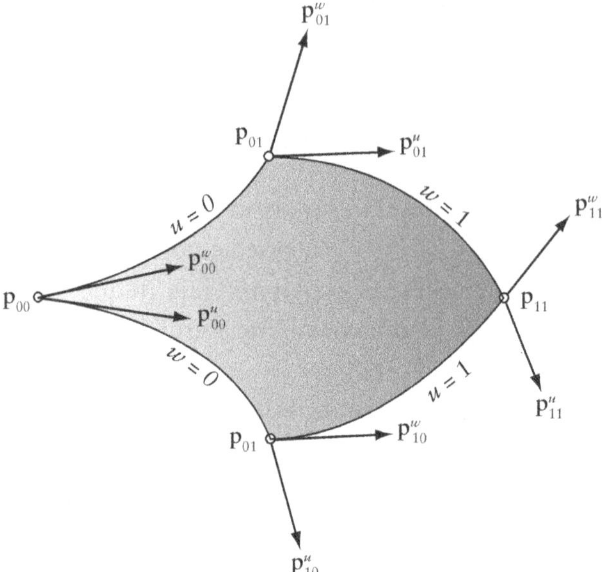

Figure 11.12 Boundary vectors defining a surface patch

11.6 Local Properties of a Surface

Just as two local properties, curvature and torsion, determine a unique curve, in a similar way local properties called the first and second fundamental forms determine a unique surface. We denote these as Form I and Form II. They are presented here without derivation or proof. If $\mathbf{p}(u, w)$ represents a parametric surface, then its first fundamental form is

$$\text{Form I:} \quad d\mathbf{p} \bullet d\mathbf{p} = Edu^2 + 2Fdudw + Gdw^2, \tag{11.68}$$

where

$$E = \mathbf{p}^u \bullet \mathbf{p}^u, \tag{11.69}$$

$$F = \mathbf{p}^u \bullet \mathbf{p}^w, \tag{11.70}$$

$$G = \mathbf{p}^w \bullet \mathbf{p}^w. \tag{11.71}$$

E, F, and G are known as the coefficients of the first fundamental form. In the metric theory of surfaces, the first fundamental form arises when calculating the arc length of a curve on a surface.

If $\mathbf{n}(u, w)$ is the unit normal to a surface at a point $\mathbf{p}(u, w)$, then the second fundamental form is

$$\text{Form II:} -d\mathbf{p}(u, w) \bullet d\mathbf{n}(u, w) = Ldu^2 + 2Mdudw + Ndw^2, \tag{11.72}$$

where

$$L = -\mathbf{p}^u \bullet \mathbf{n}^u, \tag{11.73}$$

$$M = -1/2(\mathbf{p}^u \bullet \mathbf{n}^w + \mathbf{p}^w \bullet \mathbf{n}^u), \tag{11.74}$$

$$N = -\mathbf{p}^{w} \bullet \mathbf{n}^{w} \tag{11.75}$$

and where $\mathbf{p}^{u} = \partial \mathbf{p}(u, w)/\partial u$, etc. L, M, and N are the coefficients of the second fundamental form.

Note that $\mathbf{n}^{u}$ and $\mathbf{n}^{w}$ are perpendicular to $\mathbf{n}$. Then, because $\mathbf{p}^{u}$ and $\mathbf{p}^{w}$ are perpendicular to $\mathbf{n}$ for all u, w, we can derive alternative expressions for L, M, and N:

$$L = \mathbf{p}^{uu} \bullet \mathbf{n}, \tag{11.76}$$

$$M = \mathbf{p}^{uw} \bullet \mathbf{n}, \tag{11.77}$$

$$N = \mathbf{p}^{ww} \bullet \mathbf{n}. \tag{11.78}$$

It turns out that the coefficients of these forms are not invariant under a parameter transformation. It is possible to demonstrate the use of these coefficients of the first fundamental form to calculate the arc length of curves on surfaces, angles, and surface areas.

To review, the unit normal at any point on a bicubic patch is

$$\mathbf{n} = \frac{\mathbf{p}^{u} \times \mathbf{p}^{w}}{\left|\mathbf{p}^{u} \times \mathbf{p}^{w}\right|}. \tag{11.79}$$

If $\mathbf{p}^{u}$ and $\mathbf{p}^{w}$ are perpendicular to each other, then $\mathbf{p}^{u}$, $\mathbf{p}^{w}$, and $\mathbf{n}$ form a local orthogonal basis or trihedron at the point. Or we can easily construct one by any number of other ways. For example, $\mathbf{p}^{u}$, $\mathbf{q}$, and $\mathbf{n}$ form a basis if we define $\mathbf{q} = \mathbf{n} \times \mathbf{p}^{u}$. We find

$$\mathbf{q}(s, t) = \mathbf{p}_i + s\mathbf{p}_i^{u} + t\mathbf{p}_i^{w}. \tag{11.80}$$

The equation of a plane tangent to a point $\mathbf{p}(u_i, w_i)$ on a surface is

$$\mathbf{q}(s, t) = \mathbf{p}(u_i, w_i) + s\mathbf{p}_i^{u} + t\mathbf{p}_i^{w}, \tag{11.81}$$

or expressed as a determinant,

$$\begin{vmatrix} x - x_i & x_i^{u} & x_i^{w} \\ y - y_i & y_i^{u} & y_i^{w} \\ z - z_i & z_i^{u} & z_i^{w} \end{vmatrix} = 0, \tag{11.82}$$

where x, y, and z are the coordinates of any point $\mathbf{q}(s, t)$ on the tangent plane; x_i, y_i, and z_i are the coordinates of the point $\mathbf{p}(u_i, w_i)$ on the surface to which the plane is tangent; and x_i^{u}, y_i^{u}, and z_i^{u} and x_i^{w}, y_i^{w}, and z_i^{w} are components of the parametric tangent vectors $\mathbf{p}^{u}(u_i, w_i)$ and $\mathbf{p}^{w}(u_i, w_i)$, respectively, at $\mathbf{p}(u_i, w_i)$.

As a vector expression, this determinant becomes

$$(\mathbf{q}-\mathbf{p})\bullet(\mathbf{p}^u \times \mathbf{p}^w)=0. \tag{11.83}$$

Figure 11.13 illustrates this situation. Because $\mathbf{p}^u(u_i, w_i)$ and $\mathbf{p}^w(u_i, w_i)$ lie in a plane, their vector product is perpendicular to the plane, defining the direction of the unit normal $\mathbf{n}$. The tangent plane must be perpendicular to the normal, so that any point $\mathbf{q}$ in it defines the vector $\mathbf{q} - \mathbf{p}$. Therefore, Equation 11.83 must hold.

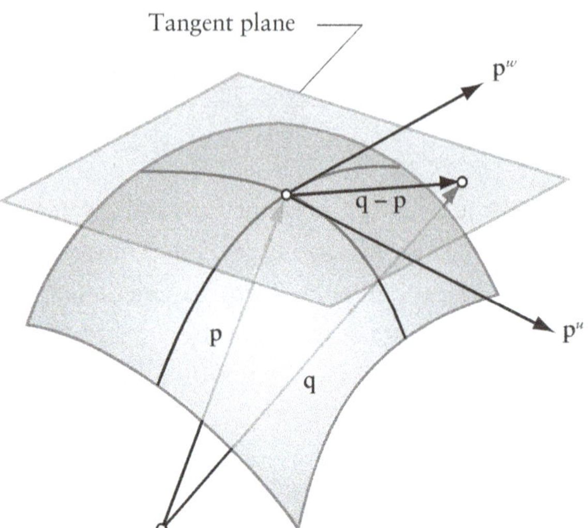

Figure 11.13 Tangent plane

The normal at a point on a surface is perpendicular to both parametric tangents at that point. In fact, at an ordinary point on a surface, a remarkable property of the geometry of a surface is that the tangent lines of all the curves lying on it and passing through that common point lie in a common plane. This is, indeed, the tangent plane. At extraordinary points on some surfaces, although they do possess a unique normal, the tangent plane can also intersect the surface. To study these, see more advanced texts on differential geometry.

11.7 Global Properties of a Surface

Global properties are those that depend on the overall characteristics of a geometric object. Here we will consider surface area, volume, and certain characteristic tests for surfaces.

The bicubic patch and other parametric forms are well adapted to computing geometric properties, since we do not have to compute explicit points. To compute surface area, we use an elementary property of vectors (shown in Figure 11.14):

$$\begin{aligned} dA &= |\,\mathbf{n}(u, w)\,|\, du dw \\ &= f_1(u, w) du dw, \end{aligned} \tag{11.84}$$

where dA is the scalar element of area and $\mathbf{n}(u, w)$ is a vector function defining the patch normals.

Therefore,

$$A = \int_0^1 \int_0^1 f_1(u, w)\, du\, dw. \tag{11.85}$$

Finally, using Gaussian quadrature, we obtain

$$A = \sum_{i=0}^{n} \sum_{j=0}^{n} g_i h_j f_1(u_j, w_i), \tag{11.86}$$

where g_i and h_j are the weight values associated with a specific n-point formula.

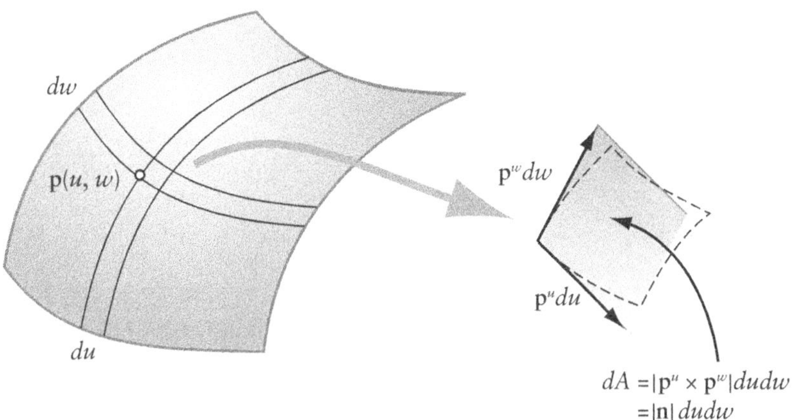

Figure 11.14 Surface area

For the volume of a closed region, we have from vector calculus

$$\begin{aligned} dV &= \frac{1}{3}[\mathbf{p}(u,w) \bullet \mathbf{n}(u,w)]\, du dw \\ &= f_2(u,w) du dw, \end{aligned} \tag{11.87}$$

where dV is the scalar element of volume, $\mathbf{p}$ is the position vector, and $\mathbf{n}$ is the normal vector (Figure 11.15).

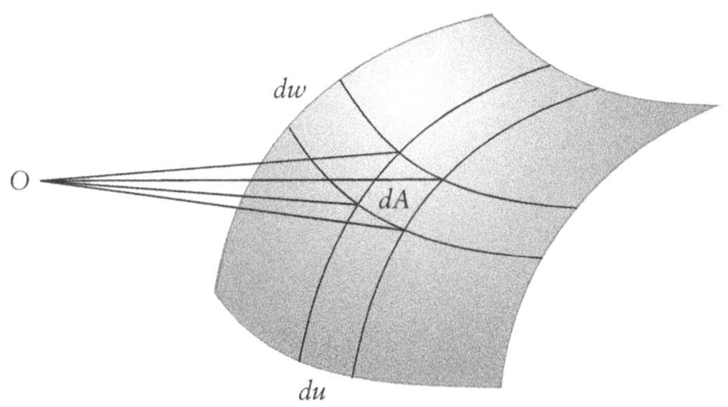

Figure 11.15 Volume

The factor of 1/3 arises from the solid geometry formula for a pyramid (that is, the volume is 1/3 the product of the base area and the perpendicular height). The scalar product of the vectors satisfies the requirement of perpendicularity between the base and height. So we obtain

$$V = \int_0^1 \int_0^1 f_2(u, w)\, du\, dw, \tag{11.88}$$

and we may again apply quadrature as we did to obtain Equation 11.86 from Equation 11.85.

Here are two simple tests we can apply to surfaces to determine if they possess special characteristics. A surface is planar if $\kappa_n = 0$ at all points, or it is spherical if $\kappa_n = \text{constant}$. Interestingly, a surface is developable if its Gaussian curvature K is zero everywhere, which means that

$$\begin{aligned} K &= \kappa_1 \kappa_2 \\ &= \frac{LN - M^2}{EG - F^2} = 0, \end{aligned} \tag{11.89}$$

or

$$LN - M^2 = 0. \tag{11.90}$$

Here we will stop, although there is much more to be said about the properties of both curves and surfaces. The subject of surfaces is deep and requires the application of differential geometry. In fact, differential geometry was developed and refined as a way to study curves and surfaces. Here we have only scratched the surface, so to speak, with the objective of showing the importance of vectors. The ideas revealed here are central to geometric and 3D modeling.

12 Spatial Relationships

Local and global geometric properties tell us a lot about a geometric object itself. But how about its relationship to other geometric objects sharing the same space. How far apart is one from another? Do they intersect? Will they collide if moved about via motion transformations? These questions and similar ones frequently arise in the world of CAD/CAM, animation and game design. This chapter introduces vector descriptions and solutions for some of these spatial relationship problems.

12.1 Distance

Distance is a measure of how far apart two geometric objects are. In physics or everyday usage, distance may refer to a physical length. In most cases, "distance from A to B" is interchangeable with "distance from B to A." In mathematics, a distance function or metric is a generalization of the concept of physical distance. A metric is a function that behaves according to a specific set of rules, and is a way of describing what it means for elements of some space to be "close to" or "far away from" each other.

A metric is a mathematical function that defines distance. The Euclidean metric for a Cartesian coordinate system is the most familiar and most often used in everyday geometry and engineering. It looks like this:

$$d = \sqrt{(x_2 - x_1)^2 + (y_2 - y_1)^2 + (z_2 - z_1)^2}, \tag{12.1}$$

which is the distance between two points in a three-dimensional Cartesian coordinate system erected in Euclidean space. Note that distance is always positive. Familiar, indeed!

In the context of geometric and 3D modeling, the magnitude of a vector usually refers to length or, equivalently, distance. In other contexts, engineering and physics, for example, the magnitude of a vector might signify a force, velocity, acceleration, or some other phenomenon that has magnitude and direction attached to it. Earlier we discovered that the length of a vector is the square root of the dot product with itself:

$$|\mathbf{p}| = \sqrt{\mathbf{p} \bullet \mathbf{p}}. \tag{12.2}$$

Coordinate systems are possible in curved spaces. For example, we can construct coordinates on the closed surface of a sphere (latitude and longitude are one way), along a curve in space, and on open curved surfaces.

The distance or length of shortest paths between points in these systems is usually much more complicated.

Consider the distance problems posed in Figure 12.1. First, the minimum distance between two arbitrary points in space $\mathbf{p}_1$ and $\mathbf{p}_2$ is

$$d_{\min} = |\mathbf{p}_2 - \mathbf{p}_1| = \sqrt{(\mathbf{p}_2 - \mathbf{p}_1) \bullet (\mathbf{p}_2 - \mathbf{p}_1)}. \tag{12.3}$$

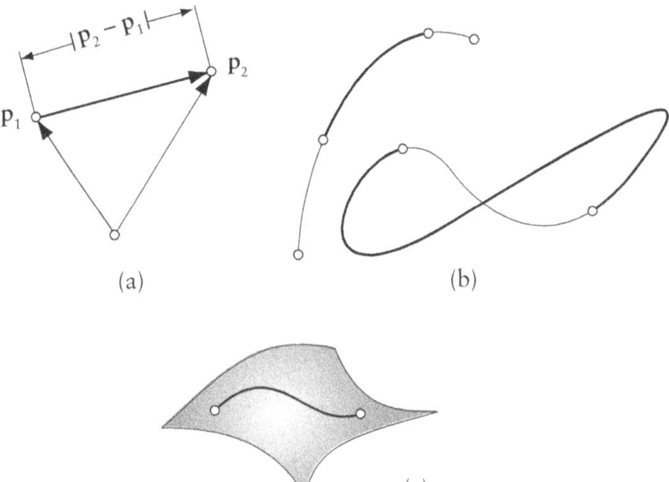

Figure 12.1 Minimum distance between two points

To find the path length between two points $\mathbf{p}(u_1)$ and $\mathbf{p}(u_2)$ along a curve, we first ask if the curve is an open segment or a closed loop. There is only one path between two points on a curve segment, and so the problem is to find the arc length between them. If the curve is closed, then there are two possible candidates, one of which is the minimum distance. One approach is to sum up chord lengths between a set of intermediate points. Another is to sum up a series of circular arcs using a series of three intermediate points to calculate arc lengths.

To find the minimum-length path between two points on a surface requires finding the shortest curve segment that lies in the surface and passes through both points. This is not a trivial problem and in some cases requires the mathematics of geodesics. Often there is no direct solution, and we must compute by iterative methods and trial and error, and there may be more than one shortest path.

12.2 Minimum Distances

Now let's look at some minimum-distance problems involving points and how vectors can help us solve them. In some of these there is no direct solution, and complex numerical methods are necessary, which are not discussed here. What is discussed is the vector definition of a specific problem and how it often leads to a solution.

Minimum Distance Between a Point and a Curve

The minimum distance from a given point **q** to a curve **p**(u) is determined by finding a vector **p** – **q** from the point to the curve that is perpendicular to the tangent vector $\mathbf{p}^u$ at **p**. Figure 12.2 shows the vector geometry.

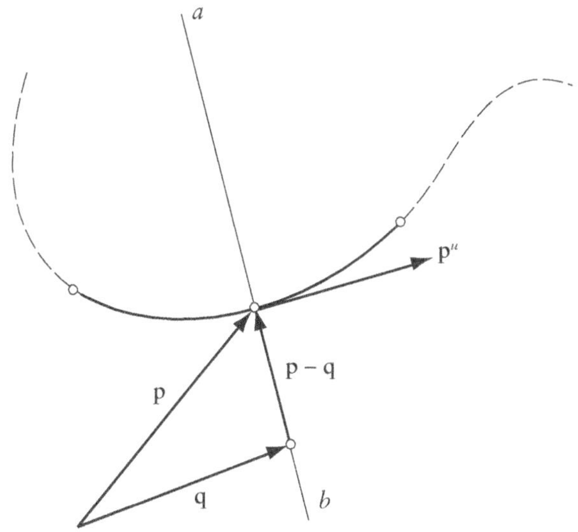

Figure 12.2 Minimum distance between a point and a curve

Mathematically, we express the required conditions as

$$d_{\min} = |\,\mathbf{p}-\mathbf{q}\,|, \tag{12.4}$$

when

$$(\mathbf{p}-\mathbf{q})\bullet\mathbf{p}^u = 0. \tag{12.5}$$

For a parametric cubic curve, this equation produces a quintic polynomial in u, one value (root) of which will produce the minimum distance via Equation 12.5. Numerical methods are usually required. Approximating a curve with straight-line segments is another option.

Minimum Distance Between a Point and a Plane

An important special case of finding the minimum distance between a point and a surface is finding the minimum distance between a point and a plane. Let the point be denoted by **q** and the plane by $k\mathbf{n}$, where **n** is the unit normal vector to the plane from the origin (Figure 12.3). In addition, let **p** denote the point on the plane closest to **q**. Then we again recognize that $d_{\min} = |\,\mathbf{p}-\mathbf{q}\,|$, where **p** must satisfy

$$(\mathbf{p}-\mathbf{q})\times\mathbf{n} = 0. \tag{12.6}$$

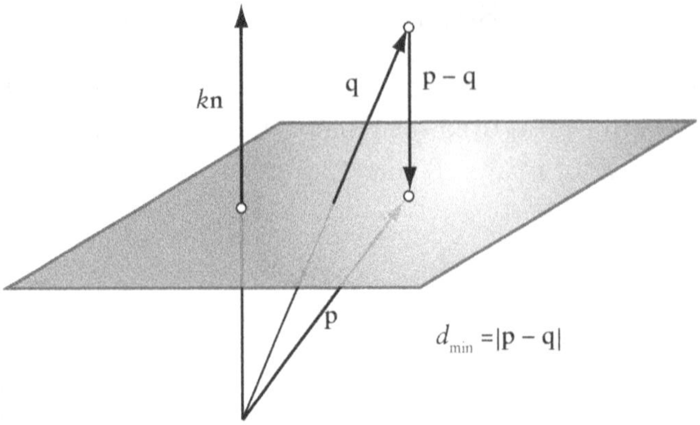

Figure 12.3 Minimum distance between a point and a plane

Equation 12.6 merely asserts that the vector $(\mathbf{p} - \mathbf{q})$ must be parallel to the normal to the plane, $\mathbf{n}$. The point $\mathbf{p}$ must also satisfy the equation of the plane $Ax + By + Cz + D = 0$ or some other equivalent constraint, such as

$$(\mathbf{p}-\mathbf{q})\bullet(\mathbf{p}-k\mathbf{n}) = 0. \tag{12.7}$$

Minimum Distance Between a Point and a Surface

Figure 12.4 shows the more general problem. Here we want to find the minimum distance between a point and a nonplanar surface. Again, let the point be denoted by $\mathbf{q}$ and the point on the surface closest to it by $\mathbf{p}(u, w)$. Then, clearly, $d_{\text{min}} = |\,\mathbf{p}-\mathbf{q}\,|$. The vector $(\mathbf{p} - \mathbf{q})$ must be in the direction of the surface normal at $\mathbf{p}$, which means that $\mathbf{p}$ must satisfy

$$(\mathbf{p}-\mathbf{q})\times(\mathbf{p}^u - \mathbf{q}^w) = 0, \tag{12.8}$$

or

$$(\mathbf{p}-\mathbf{q}) = a(\mathbf{p}^u - \mathbf{q}^w). \tag{12.9}$$

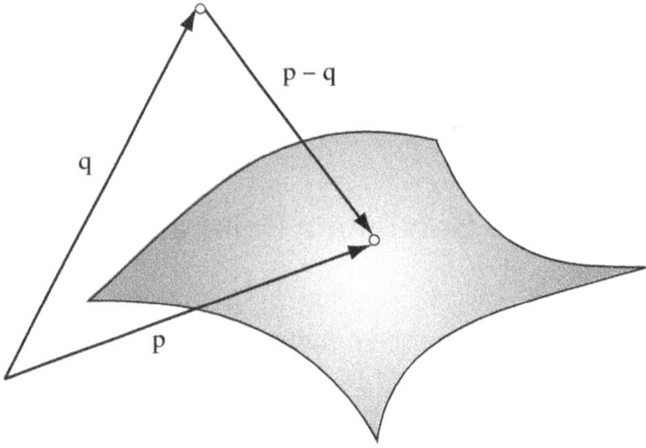

Figure 12.4 Minimum distance between a point and a surface

As with points and curves, a numerical analysis method must be employed to determine $\mathbf{p}$. Notice that unless the surface is closed, there may not be a normal in the interval $u, w \in [0, 1]$ or, if the surface is composite, within its parametric domain. Again, you may have to check for a concave relationship producing a maximum that also satisfies Equation 12.8 or 12.9. Or there may be more than one normal and more than one closest point. Remember to check boundary curves and patch corner points. A final possibility is a locus of equidistant closest points describing a curve on the surface. Sophisticated modeling systems are capable of detecting and resolving these possibilities.

Minimum Distance Between Two Curves

Given two parametric curves $\mathbf{p}(s)$ and $\mathbf{q}(t)$, we want to find the minimum distance between them (Figure 12.5). Let $\mathbf{p}$ and $\mathbf{q}$ denote corresponding points of closest approach on the respective curves. Then $d_{\min} = |\mathbf{p} - \mathbf{q}|$, and these points satisfy the following conditions:

$$(\mathbf{p} - \mathbf{q}) \bullet \mathbf{p}^s = 0 \tag{12.10}$$

and

$$(\mathbf{p} - \mathbf{q}) \bullet \mathbf{q}^t = 0. \tag{12.11}$$

These equations produce two nonlinear simultaneous equations to be solved. Only those solution pairs (s, t) in the interval $s, t \in [0, 1]$ are of interest. Notice that there can be more than one pair of points that satisfy Equations 12.10 and 12.11. We must compare each solution with the others to find the minimum; watch for local maximums.

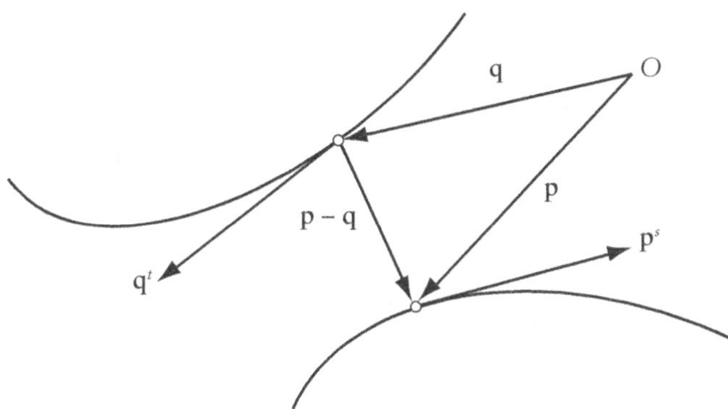

Figure 12.5 Minimum distance between two curves

Minimum Distance Between a Curve and a Plane

Consider the special problem of finding the minimum distance between a curve and a plane (Figure 12.6). Denote a point on the curve $\mathbf{p}(u)$ as $\mathbf{p}$; the plane as $k_n\mathbf{n}$, where $\mathbf{n}$ is the normal to the plane from the origin; and $\mathbf{q}$, the point on the plane closest to the curve. Then, as usual, the minimum distance is $d_{\min} = |\mathbf{p} - \mathbf{q}|$, and these points satisfy the following conditions:

$$(\mathbf{p} - \mathbf{q}) \bullet \mathbf{p}^u = 0 \tag{12.12}$$

and

$$(\mathbf{p}-\mathbf{q})\times\mathbf{n}=0. \tag{12.13}$$

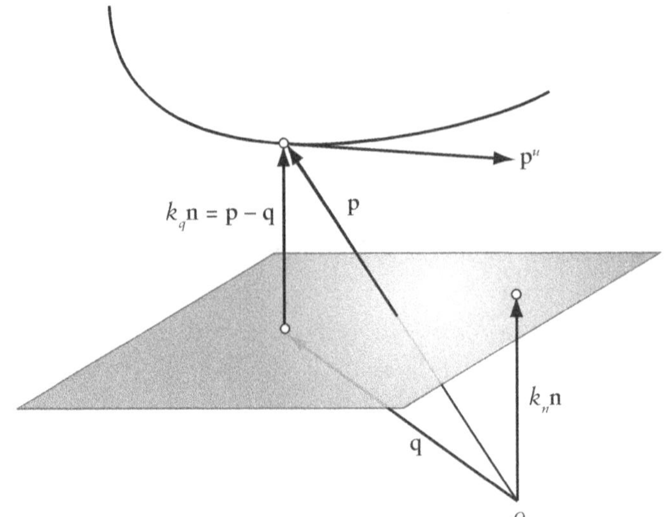

Figure 12.6 Minimum distance between a curve and a plane

One approach is to express Equation 12.12 as $k_q\mathbf{n}\bullet\mathbf{p}^u=0$. Since k_q is a scalar, we immediately obtain $\mathbf{n}\bullet\mathbf{p}^u=0$. Performing vector arithmetic results in the quadratic equation $au^2+bu+c=0$, where the constants a, b, c consist of algebraic coefficients of **p** and the components of **n** (that is, n_x, n_y, n_z). It is possible that both roots of the quadratic are in the interval $u\in[0,1]$. If so, proceed with both resulting candidate values of **p** to the next step, computing **q** from Equation 12.13. Then select the smallest value of $|\mathbf{p}-\mathbf{q}|$ as the correct solution. Be aware that the following conditions are possible: There are no roots in the interval, and one of the endpoints is closest; the curve and the plane intersect; or the curve and plane are parallel.

Minimum Distance Between a Curve and a Surface

Figure 12.7 illustrates the more general problem of finding the minimum distance between a cubic curve and a bicubic surface patch. The vector **p** − **q** must satisfy the following conditions if it is a minimum-distance candidate:

$$(\mathbf{p}-\mathbf{q})\times(\mathbf{p}^u\times\mathbf{p}^w)=0 \tag{12.14}$$

and

$$(\mathbf{p}-\mathbf{q})\bullet\mathbf{q}^t=0. \tag{12.15}$$

These two equations generate a system of simultaneous, nonlinear equations for whose solution, happily, numerical methods are available.

Check for local maximums, patch corner points, and boundary curves and curve endpoints.

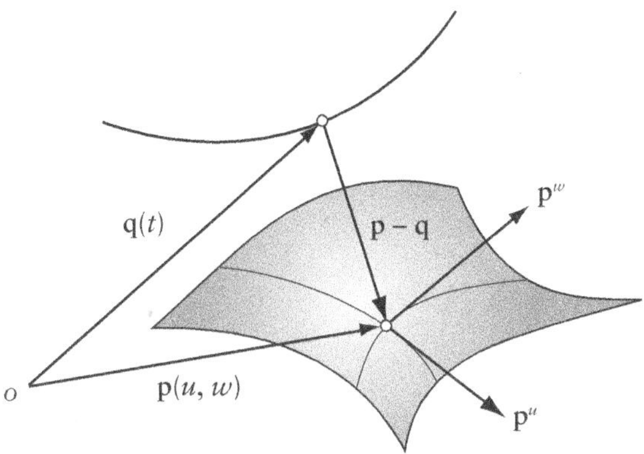

Figure 12.7 Minimum distance between a curve and a surface

Minimum Distance Between Two Surfaces

The minimum distance between two surface patches $\mathbf{p}(u, w)$ and $\mathbf{q}(s, t)$ is $|\mathbf{p}-\mathbf{q}|$ when these vectors satisfy the following two conditions:

$$(\mathbf{p}-\mathbf{q})\times(\mathbf{p}^u \times \mathbf{p}^w) = 0 \tag{12.16}$$

and

$$(\mathbf{p}-\mathbf{q})\times(\mathbf{q}^s \times \mathbf{q}^t) = 0. \tag{12.17}$$

Figure 12.8 illustrates the vector geometry describing this problem.

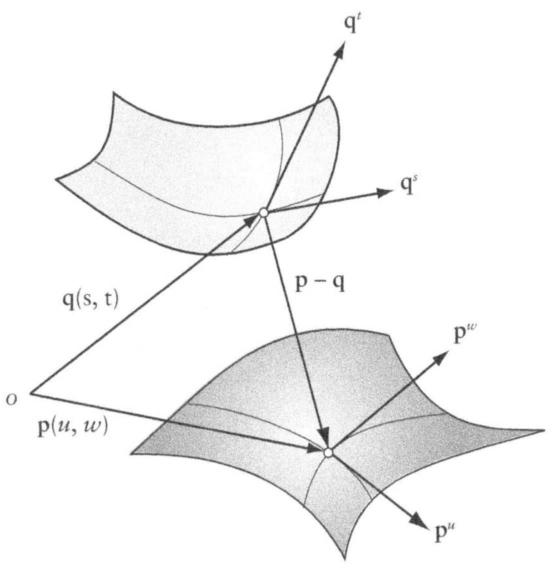

Figure 12.8 Minimum distance between two surfaces

12.3 Line Intersections

One of the most important computational tasks of modeling is determining whether or not two geometric objects intersect and what is the geometry of the intersection itself. So if two objects do intersect, we must be able to mathematically describe their intersection. As the number and the complexity of objects increase, computational complexity also increases. We begin with the simple problem of finding the points of intersection between a straight line and other geometric objects.

Intersection of Two Lines

Two straight-line segments in three-dimensional space may intersect in two different ways, if they intersect at all. They may intersect at a single point, as at **r** in Figure 12.9, or they may be collinear (that is, overlap either partially or completely).

Let one line be given by $\mathbf{p}(u) = \mathbf{a} + u\mathbf{b}$ and the other by $\mathbf{q}(w) = \mathbf{c} + w\mathbf{d}$, where $u, w \in [0, 1]$. At their point of intersection, $\mathbf{p}(u) = \mathbf{q}(w) = \mathbf{r}$, or

$$\mathbf{a} + u\mathbf{b} = \mathbf{c} + w\mathbf{d}. \tag{12.18}$$

Equation 12.18 represents three linear equations, one for each component or coordinate direction, and two unknowns, u and w. The system is clearly overconstrained. Solve for u and w using two of the equations, say, $a_x + ub_x = c_x + wd_x$ and $a_y + ub_y = c_y + wd_y$. Then use the third equation, $a_z + ub_z = c_z + wd_z$, to verify the results. Finally, use either member of the intersection solution pairs (u_i, w_i) to find **r**; for example, $\mathbf{r} = \mathbf{a} + u_i\mathbf{b}$.

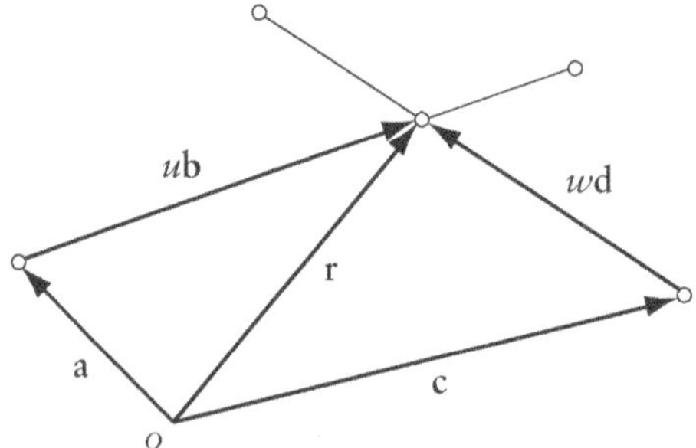

Figure 12.9 Intersection of two straight lines

If we use some fundamental properties of vectors, we can compute u and w directly as follows: To solve Equation 12.18 for u, take the indicated vector and scalar products

$$(\mathbf{c} \times \mathbf{d}) \bullet (\mathbf{a} + u\mathbf{b}) = (\mathbf{c} \times \mathbf{d}) \bullet (\mathbf{c} + w\mathbf{d}). \tag{12.19}$$

Since $(\mathbf{c} \times \mathbf{d})$ is perpendicular to both **c** and **d**, the right side of this equation becomes zero, and we obtain for u

$$u = \frac{(\mathbf{c} \times \mathbf{d}) \bullet \mathbf{a}}{(\mathbf{c} \times \mathbf{d}) \bullet \mathbf{b}}. \tag{12.20}$$

A similar approach yields

$$w = \frac{(\mathbf{a} \times \mathbf{b}) \bullet \mathbf{c}}{(\mathbf{a} \times \mathbf{b}) \bullet \mathbf{d}}. \tag{12.21}$$

Intersection of a Line and a Plane

Consider the intersection of a straight line (or, equivalently, a ray) with an unbounded plane (Figure 12.10). The vector equation $\mathbf{p}(u, w) = \mathbf{a} + u\mathbf{b} + w\mathbf{c}$ defines points on the plane, and $\mathbf{q}(t) = \mathbf{d} + t\mathbf{e}$ defines points on the line or ray. Their point of intersection is the vector $\mathbf{r}$. If they intersect, then $\mathbf{p}(u, w) = \mathbf{q}(t) = \mathbf{r}$, or

$$\mathbf{a} + u\mathbf{b} + w\mathbf{c} = \mathbf{d} + t\mathbf{e}. \tag{12.22}$$

Equation 12.22 represents three linear equations and three unknowns. Using some fundamental vector properties, isolate, in turn, u, w, and t. For t, apply $\mathbf{b} \times \mathbf{c}$ as follows:

$$(\mathbf{b} \times \mathbf{c}) \bullet (\mathbf{a} + u\mathbf{b} + w\mathbf{c}) = (\mathbf{b} \times \mathbf{c}) \bullet (\mathbf{d} + t\mathbf{e}). \tag{12.23}$$

Because $\mathbf{b} \times \mathbf{c}$ is perpendicular to both $\mathbf{b}$ and $\mathbf{c}$, then

$$(\mathbf{b} \times \mathbf{c}) \bullet \mathbf{a} = (\mathbf{b} \times \mathbf{c}) \bullet \mathbf{d} + (t\mathbf{b} \times \mathbf{c}) \bullet \mathbf{e} \tag{12.24}$$

and

$$t = \frac{(\mathbf{b} \times \mathbf{c}) \bullet \mathbf{a} - (\mathbf{b} \times \mathbf{c}) \bullet \mathbf{d}}{(\mathbf{b} \times \mathbf{c}) \bullet \mathbf{e}}. \tag{12.25}$$

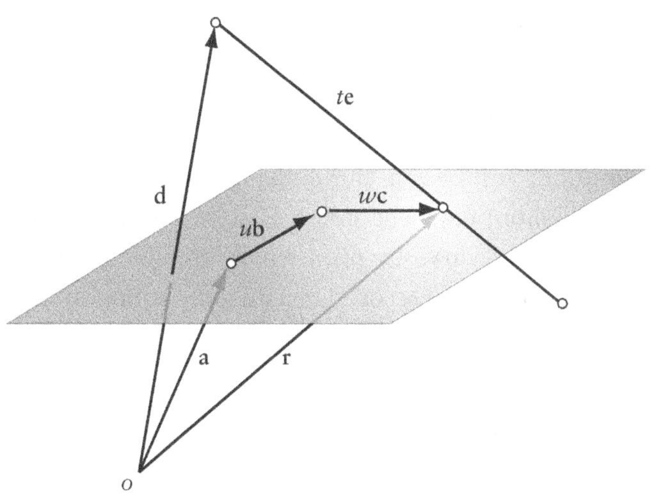

Figure 12.10 Intersection of a straight line and a plane

There are similar expressions for u and w:

$$u = \frac{(\mathbf{c}\times\mathbf{e})\bullet\mathbf{d}-(\mathbf{c}\times\mathbf{e})\bullet\mathbf{a}}{(\mathbf{c}\times\mathbf{e})\bullet\mathbf{b}}, \tag{12.26}$$

$$w = \frac{(\mathbf{b}\times\mathbf{e})\bullet\mathbf{d}-(\mathbf{b}\times\mathbf{e})\bullet\mathbf{a}}{(\mathbf{b}\times\mathbf{e})\bullet\mathbf{c}}. \tag{12.27}$$

If this plane contains a three-sided polygon, one of a large set of polygons that together approximate the curved surface of a 3D model, then we can determine if a ray intersects the plane within the boundaries of the triangular polygon. Here is how we proceed.

First, you may want to review Chapter 7 on barycentric coordinates. Let $\mathbf{p}_{\text{int}}$ denote the vector representing the point of intersection of the ray with the plane of the triangle. We can write $\mathbf{p}_{\text{int}}$ in terms of its barycentric coordinates with respect to the triangle:

$$\mathbf{p}_{\text{int}} = b_1\mathbf{v}_1 + b_2\mathbf{v}_2 + b_3\mathbf{v}_3, \tag{12.28}$$

where $\mathbf{v}_1$, $\mathbf{v}_2$, and $\mathbf{v}_3$ are vectors defining the vertex points of the triangle.

We solve Equation 12.28 to find b_1, b_2, and b_3. If the values of the barycentric coordinates are each less than 1, then the point lies inside the triangle. That is, the ray intersects the plane within the triangle defined by $\mathbf{v}_1$, $\mathbf{v}_2$, and $\mathbf{v}_3$.

To find the ray's angle of intersection, θ, with the plane of the triangle, we compute the normal to the plane (from Figure 12.10):

$$\mathbf{n} = \mathbf{b}\times\mathbf{c}, \tag{12.29}$$

and θ from

$$\theta = \cos^{-1}\frac{\mathbf{n}\bullet\mathbf{e}}{|\mathbf{n}||\mathbf{e}|}. \tag{12.30}$$

Together, this angle and the barycentric coordinates of the intersection allow a rendering program to compute reflection effects and color shading for every three-sided polygonal facet of a curved surface in a computer graphics–generated scene.

Intersection of a Line and a Curve

A straight line may intersect a curve, composite curve, or closed curve (Figure 12.11) at more than one point. Denote the straight-line segment by $\mathbf{q}(t) = \mathbf{a} + t\mathbf{b}$ and the curve by $\mathbf{p}(u)$, which may be a cubic function, a Bézier or B-spline curve, and so on. Then an intersection occurs anywhere $\mathbf{p}(u) = \mathbf{q}(t) = \mathbf{r}$, or

$$\mathbf{p}(u) - \mathbf{q}(t) = 0. \tag{12.31}$$

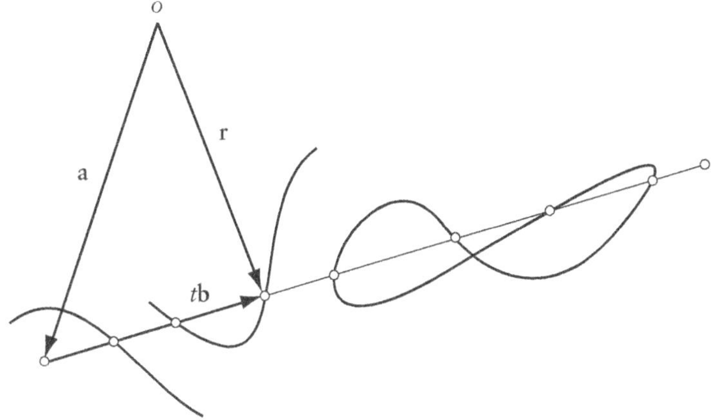

Figure 12.11 Intersection of a straight line and a curve

If $\mathbf{p}(u)$ is a composite curve comprising piecewise parametric cubic segments, we obtain three nonlinear equations in two unknowns for each segment. This is an overconstrained system. For any segment, Equation 12.31 takes the general form

$$\mathbf{c}u^3 + \mathbf{d}u^2 + \mathbf{e}u + \mathbf{f} = \mathbf{a} + \mathbf{b}t. \tag{12.32}$$

Again, some simple vector arithmetic lets us isolate u:

$$(\mathbf{a} \times \mathbf{b}) \bullet (\mathbf{c}u^3 + \mathbf{d}u^2 + \mathbf{e}u + \mathbf{f}) = 0. \tag{12.33}$$

The roots of this equation are found by numerical analysis methods. Notice that we cannot isolate t, since that would cause a loss of information because u is a cubic function with three roots.

Intersection of a Line and a Surface

Finally, consider the intersection of a straight-line segment with a surface, as Figure 12.12 illustrates. Let points on the surface be $\mathbf{p}(u, w)$ and on the straight-line segment be $\mathbf{q}(t) = \mathbf{a} + \mathbf{b}t$. Points of intersection occur when simultaneous sets of u, w, and t satisfy

$$\mathbf{p}(u, w) - \mathbf{q}(t) = 0, \tag{12.34}$$

which represents three simultaneous, nonlinear equations in three unknowns. Once again, the solution requires numerical analysis methods. Of course, we must also anticipate multiple intersections or pathological conditions. How is the solution simplified by rotating and translating the line and surface together so that the line is collinear with the z axis?

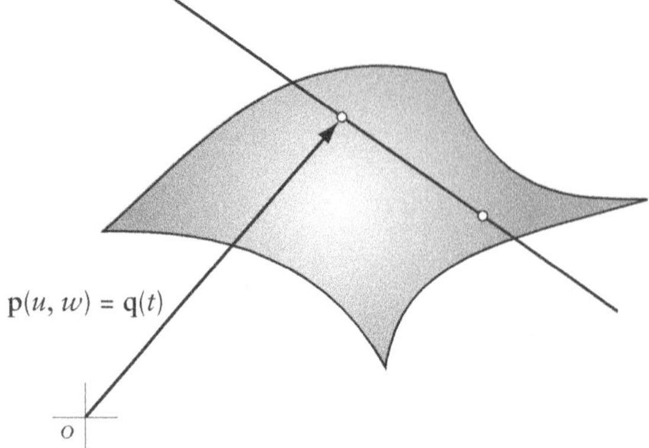

Figure 12.12 Intersection of a straight line and a surface

12.4 Plane Intersections

Computing plane intersections with geometric objects is another key requirement of 3D modeling and CAD/CAM applications, as well as many others. We can obtain most conventional cross sections of 3D objects by computing the appropriate plane intersections.

Intersection of Two Planes

To begin our study of plane intersections, consider the intersection of two planes $\mathbf{p}(u, w)$ and $\mathbf{q}(s, t)$ (Figure 12.13). Assume for the moment that the planes are bounded by $u, w, s, t \in [0, 1]$. Furthermore, note that the expected intersection is a straight line. This is revealed mathematically by observing that there are three equations from the intersection equation $\mathbf{p}(u, w) - \mathbf{q}(s, t) = 0$ and four unknowns or variables. The extra degree of freedom implies that the intersection is a curve (a straight line in the case at hand), and so we must introduce an additional constraint (see below). The expanded intersection equation looks like

$$\mathbf{a} + \mathbf{b}u + \mathbf{c}w = \mathbf{d} + \mathbf{e}s + \mathbf{f}t. \tag{12.35}$$

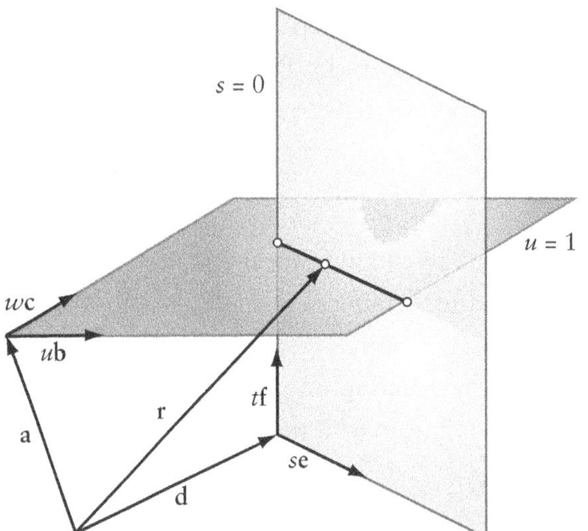

Figure 12.13 Intersection of two planes

We may now arbitrarily fix as constant one of the variables, either u, w, s, or t. This furnishes an additional equation or constraint (for example, $u - u_i$, where u_i is a constant). Then we employ techniques developed in the preceding subsections to obtain one point on the intersection. By incrementing u_i by some Δu, we find a series of points sufficient to define the intersection.

When the two intersecting planes are bounded, as just described, we can expedite the solution by recognizing, first, that the line of intersection is bounded and, second, that the endpoints of this line each lie on a boundary line of one of the planes (for example, $u = 0$, $u = 1$, and so on). This gives us a clue about how to proceed with a solution.

If the planes are not parallel, coincident, or nonintersecting, they intersect at a single point or along a bounded line segment. Since the two endpoints of the line are sufficient to define it, we proceed to solve Equation 12.35 for each of the additional constraint equations given by the plane boundaries, $u = 0, u = 1, w = 0, w = 1, s = 0, s = 1, t = 0, t = 1$. Once two points are found, terminate the algorithm, because the complete solution has been found.

Intersection of a Plane and a Curve

If an unbounded plane intersects a parametric curve at a point on the curve $\mathbf{p}(u)$, then the scalar product between the normal vector $k_n\mathbf{n}$ to the plane from the origin and the vector $(\mathbf{p} - k_n\mathbf{n})$ lying in the plane must equal zero (Figure 12.14). We express this as

$$(\mathbf{p} - k_n\mathbf{n}) \bullet k_n\mathbf{n} = 0. \tag{12.36}$$

Some vector algebra yields a single nonlinear equation. If the curve is defined in the interval $u \in [0, 1]$, then only roots in this interval represent valid intersections; we may find one, two, or three points of intersection for a cubic curve, or we may find none.

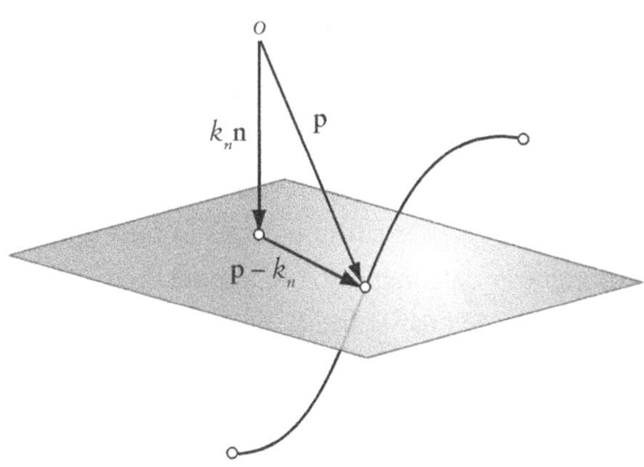

Figure 12.14 Intersection of a plane and a curve

Intersection of a Plane and a Surface

Now consider the intersection between an unbounded plane and a surface, as Figure 12.15 shows. Since we already know that a plane is also a two-parameter geometric element, we expect to derive three equations having four degrees of freedom. The extra degree of freedom manifests itself as a curve of intersection; however, since we do not require the values of the parametric variables defining the plane, we define the plane by the normal vector $k_n\mathbf{n}$ to the plane from the origin, where $\mathbf{n}$ is the unit normal vector. Then, beginning with the surface boundary curves, we search for intersection points by intersecting these curves and other carefully selected isoparametric curves with the plane, as in Equation 12.36. If the plane and object to be intersected are rotated and translated so that the plane is coincident with the $z = 0$ plane, how is the solution simplified?

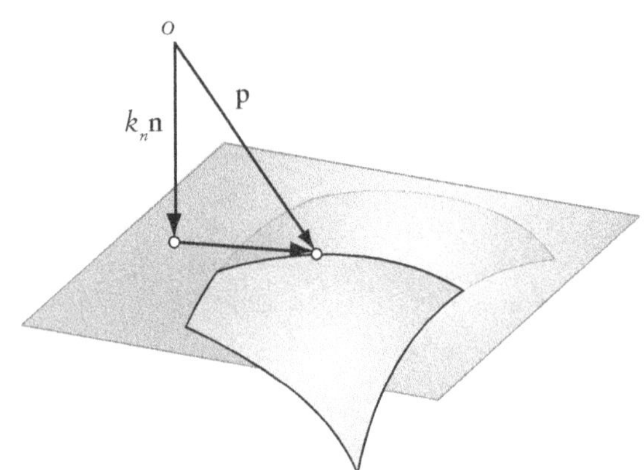

Figure 12.15 Intersection of a plane and a surface

12.5 Curve Intersections

Finding the intersection of two surfaces (not discussed in this textbook) is still a challenging problem in geometric modeling. The next two subsections are preparation for understanding and undertaking to solve the surface-surface intersection problem.

Intersection of Two Curves

Computing the intersection of two curves in three-dimensional space (Figure 12.16) requires solving for the roots of the equation $\mathbf{p}(u) - \mathbf{q}(t) = 0$, which usually offers three nonlinear equations in two unknowns. Suppress one of the component equations, say, for z, and then solve the remaining two simultaneous equations for u and t:

$$\begin{aligned} p_x(u) - q_x(t) &= 0, \\ p_y(u) - q_y(t) &= 0. \end{aligned} \tag{12.37}$$

The solution is really the intersection of the projections of the two curves onto the xy plane. Verify it by checking $p_x(u)-q_x(t)=0$. In many cases, this intersection problem is a strictly two-dimensional one, and so we express it as two equations in two unknowns.

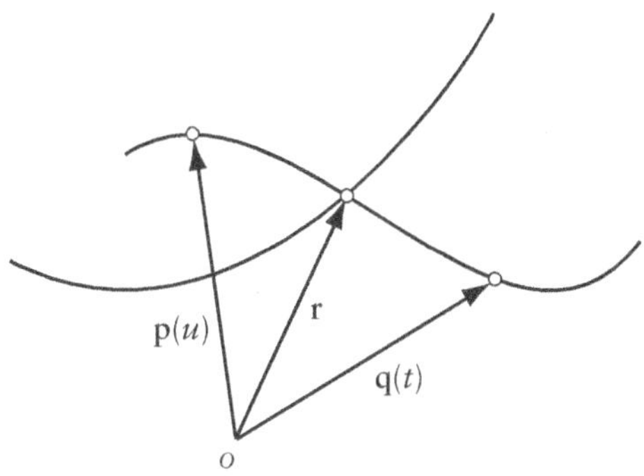

Figure 12.16 Intersection of two curves

Intersection of a Curve and a Surface

Now consider the problem of finding the intersection between a curve and a surface. This method is the basis of some algorithms for computing the intersection of two surfaces. Figure 12.17 presents the essential ingredients.

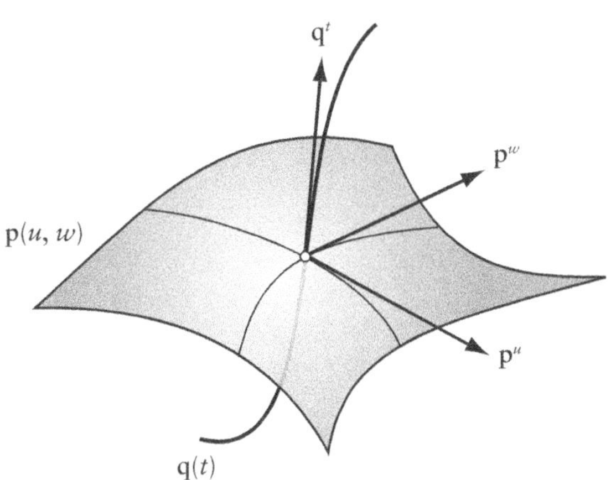

Figure 12.17 Intersection of a curve and a surface

We express the point of intersection as a fully specified system of three equations in three unknowns. Denote the curve by $\mathbf{q}(t)$ and the surface by $\mathbf{p}(u, w)$. Then express the system of equations as

$$\mathbf{r}=\mathbf{q}(t)-\mathbf{p}(u, w)=0, \tag{12.38}$$

where $\mathbf{r}$ is the minimum distance between the surface and successive points on the curve.

To begin the numerical solution, compute the first variation of this equation:

$$\mathbf{r}=\mathbf{q}^t dt-\mathbf{p}^u du-\mathbf{p}^w dw. \tag{12.39}$$

Isolate the variations in the individual parametric variables by imposing appropriate vector and scalar products on Equation 12.39; for example,

$$\mathbf{p}^u \times d\mathbf{r} = (\mathbf{p}^u \times \mathbf{q}^t)dt - (\mathbf{p}^u \times \mathbf{p}^u)du - (\mathbf{p}^u \times \mathbf{p}^w)dw. \tag{12.40}$$

Since $(\mathbf{p}^u \times \mathbf{p}^u) = 0$, we obtain

$$\mathbf{p}^u \times d\mathbf{r} = (\mathbf{p}^u \times \mathbf{q}^t)dt - (\mathbf{p}^u \times \mathbf{p}^w)dw. \tag{12.41}$$

Furthermore,

$$\mathbf{q}^t \bullet (\mathbf{p}^u \times d\mathbf{r}) = \mathbf{q}^t \bullet (\mathbf{p}^u \times \mathbf{q}^t)dt - \mathbf{q}^t \bullet (\mathbf{p}^u \times \mathbf{p}^w)dw. \tag{12.42}$$

Since $\mathbf{q}(t)$ is perpendicular to $\mathbf{p}^u \times \mathbf{q}^t$, then

$$\mathbf{q}^t \bullet \left(\mathbf{p}^u \times \mathbf{q}^t\right) = 0, \tag{12.43}$$

and therefore

$$\mathbf{q}^t \bullet \left(\mathbf{p}^u \times d\mathbf{r}\right) = -\mathbf{q}^t \bullet \left(\mathbf{p}^u \times \mathbf{p}^w\right)dw. \tag{12.44}$$

Repeat this method to isolate the variation in the other two parameters, and then construct the following equations, which support the iteration process:

$$\begin{aligned} t_{i+1} &= t_i - \frac{\mathbf{p}_i^u \bullet (\mathbf{p}_i^w \times \mathbf{r}_i)}{D}, \\ u_{i+1} &= u_i - \frac{\mathbf{p}_i^w \bullet (\mathbf{q}_i^t \times \mathbf{r}_i)}{D}, \\ w_{i+1} &= w_i - \frac{\mathbf{q}_i^t \bullet (\mathbf{p}_i^u \times \mathbf{r}_i)}{D}, \end{aligned} \tag{12.45}$$

where

$$D = \mathbf{q}_i^t \bullet (\mathbf{p}_i^u \times \mathbf{p}_i^w) = \mathbf{q}_t \bullet \mathbf{n}_i \tag{12.46}$$

and $\mathbf{n}_i$ is the normal to the surface at $\mathbf{p}_i$.

Solutions

Here you will find solutions to the exercises presented in some of the chapters. Comments are included for some.

Chapter 3 The Algebraic Vector

3.1. Given the five vectors shown in Figure 3.12, write them in component form. The figure is an orthogonal grid shown at unit intervals.

a. $\mathbf{a} = [\ 5 \quad 6\ ]$

b. $\mathbf{b} = [\ 5 \quad -5\ ]$

c. $\mathbf{c} = [\ 0 \quad 7\ ]$

d. $\mathbf{d} = [\ -7 \quad 0\ ]$

e. $\mathbf{e} = [\ -5 \quad 3\ ]$

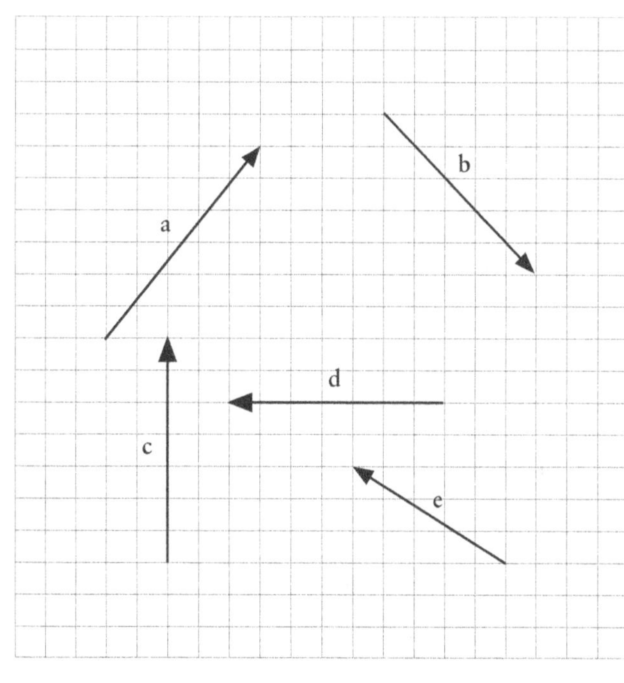

Figure 3.12 Vectors on a grid

3.2. Compute the magnitudes of the vectors given in the preceding exercise.

a. $|\mathbf{a}| = 7.81$

b. $|\mathbf{b}| = 7.07$

c. $|\mathbf{c}| = 7.00$

d. $|\mathbf{d}| = 7.00$

e. $|\mathbf{e}| = 5.83$

Magnitude and direction are the most important properties of a vector. Because a vector's magnitude is best described as a length, it is always positive. The length of a vector **a** is a scalar, denoted as $|\mathbf{a}|$ and given by

$$|\mathbf{a}| = \sqrt{a_x^2 + a_y^2 + a_z^2}.$$

This is a simple application of the Pythagorean theorem for finding the length of the diagonal of a rectangular solid.

3.3. Given that **a**, **b**, and **c** are three-component vectors, express the compact vector equation $\mathbf{a}x + \mathbf{b}y = \mathbf{c}$:

a. In expanded vector form:

Answer: $$\begin{bmatrix} a_1 \\ a_2 \\ a_3 \end{bmatrix} x + \begin{bmatrix} b_1 \\ b_2 \\ b_3 \end{bmatrix} y = \begin{bmatrix} c_1 \\ c_2 \\ c_3 \end{bmatrix}$$

b. In ordinary algebraic polynomial form:

Answer: $$\begin{aligned} a_1x + b_1y &= c_1 \\ a_2x + b_2y &= c_2 \\ a_3x + b_3y &= c_3 \end{aligned}$$

3.4. Given $\mathbf{a} = [\,-2 \quad 0 \quad 7\,]$ and $\mathbf{b} = [\,4 \quad 1 \quad 3\,]$, compute

a. $\hat{\mathbf{a}}$ Answer: $\hat{\mathbf{a}} = [\,-0.275 \quad 0 \quad 0.962\,]$

b. $\hat{\mathbf{b}}$ Answer: $\hat{\mathbf{b}} = [\,0.784 \quad 0.196 \quad 0.588\,]$

c. $\mathbf{c} = \mathbf{a} - 2\mathbf{b}$ Answer: $\mathbf{c} = [\,-10 \quad -2 \quad 1\,]$

d. $\mathbf{c} = 3\mathbf{a}$ Answer: $\mathbf{c} = [\,-6 \quad 0 \quad 21\,]$

e. $\mathbf{c} = \mathbf{a} + \mathbf{b}$ Answer: $\mathbf{c} = [\,2 \quad 1 \quad 10\,]$

A unit vector is a vector whose magnitude is equal to 1, independent of its direction, of course. We denote the unit vector in the direction of **a** as **â**, where

$$\hat{\mathbf{a}} = \frac{\mathbf{a}}{|\mathbf{a}|},$$

and its components are

$$\hat{\mathbf{a}} = \begin{bmatrix} \frac{a_x}{|\mathbf{a}|} & \frac{a_y}{|\mathbf{a}|} & \frac{a_z}{|\mathbf{a}|} \end{bmatrix}$$

3.5. Given $\mathbf{a} = [\ 6 \quad 2 \quad -5\ ]$ and $\mathbf{b} = 2\mathbf{a}$, compare the unit vectors $\hat{\mathbf{a}}$ and $\hat{\mathbf{b}}$.

Answer: $\hat{\mathbf{a}} = \hat{\mathbf{b}}$

Multiplying a vector by a scalar k produces a vector $k\mathbf{a}$, which in component form looks like this:

$$k\mathbf{a} = \begin{bmatrix} k\mathbf{a}_x & k\mathbf{a}_y & k\mathbf{a}_z \end{bmatrix}.$$

If k is positive, then **a** and $k\mathbf{a}$ are in the same direction. If k is negative, then **a** and $k\mathbf{a}$ are in opposite directions. In this exercise, $\hat{\mathbf{a}}$ and $\hat{\mathbf{b}}$ are in the same direction, since they only differ by a scalar multiple, 2. This means that they have the same direction cosines and the same unit vector.

3.6. Given $\mathbf{a} = [\ 2 \quad 3 \quad 5\ ]$ and $\mathbf{b} = [\ 6 \quad -1 \quad 3\ ]$, compute

a. $|\mathbf{a}|$ Answer: $|\mathbf{a}| = 6.16$

b. $|\mathbf{b}|$ Answer: $|\mathbf{b}| = 6.78$

c. $\mathbf{c} = \mathbf{a} + \mathbf{b}$ Answer: $\mathbf{c} = [\ 8 \quad 2 \quad 8\ ]$

d. $\mathbf{c} = \mathbf{a} - \mathbf{b}$ Answer: $\mathbf{c} = [\ -4 \quad 4 \quad 2\ ]$

e. $\mathbf{c} = 2\mathbf{a} + 3\mathbf{b}$ Answer: $\mathbf{c} = [\ 22 \quad 3 \quad 19\ ]$

3.7. Compute the following scalar products:

a. $\mathbf{i} \bullet \mathbf{i}$ Answer: $\mathbf{i} \bullet \mathbf{i} = 1$

b. $\mathbf{i} \bullet \mathbf{j}$ Answer: $\mathbf{i} \bullet \mathbf{j} = 0$

c. $\mathbf{i} \bullet \mathbf{k}$ Answer: $\mathbf{i} \bullet \mathbf{k} = 0$

d. $\mathbf{j} \bullet \mathbf{j}$ Answer: $\mathbf{j} \bullet \mathbf{j} = 1$

e. $\mathbf{j} \bullet \mathbf{k}$ Answer: $\mathbf{j} \bullet \mathbf{k} = 0$

f. $\mathbf{k} \bullet \mathbf{k}$ Answer: $\mathbf{k} \bullet \mathbf{k} = 1$

The special vectors **i**, **j**, and **k** each have a length equal to 1. The vector **i** lies along the x axis, **j** lies along the y axis, and **k** lies along the z axis, where

$$\mathbf{i} = [\ 1\ \ 0\ \ 0\ ],$$
$$\mathbf{j} = [\ 0\ \ 1\ \ 0\ ],$$
$$\mathbf{k} = [\ 0\ \ 0\ \ 1\ ].$$

The scalar product of two vectors **a** and **b** is the sum of the products of their corresponding components:

$$\mathbf{a} \bullet \mathbf{b} = a_x b_x + a_y b_y + a_z b_z.$$

3.8. Compute the magnitude and direction cosines for each of the following vectors:

a. $\mathbf{a} = [\ 3\ \ 4\ ]$

Answer: $|\mathbf{a}| = 5;\ \cos\alpha = 0.6,\ \ \cos\beta = 0.8$

b. $\mathbf{b} = [\ 0\ \ -2\ ]$

Answer: $|\mathbf{b}| = 2;\ \cos\alpha = 0,\ \ \cos\beta = -1$

c. $\mathbf{c} = [\ -3\ \ -5\ \ 0\ ]$

Answer: $|\mathbf{c}| = \sqrt{34};\ \cos\alpha = \frac{-3}{\sqrt{34}},\ \ \cos\beta = \frac{-5}{\sqrt{34}},\ \ \cos\gamma = 0$

d. $\mathbf{d} = [\ 1\ \ 4\ \ -3\ ]$

Answer: $|\mathbf{d}| = \sqrt{26};\ \cos\alpha = \frac{1}{\sqrt{26}},\ \ \cos\beta = \frac{4}{\sqrt{26}},\ \ \cos\gamma = \frac{-3}{\sqrt{26}}$

e. $\mathbf{e} = [\ x\ \ y\ \ z\ ]$

Answer: $|\mathbf{e}| = \sqrt{x^2 + y^2 + z^2};\ \cos\alpha = \frac{x}{|\mathbf{e}|},\ \ \cos\beta = \frac{y}{|\mathbf{e}|},\ \ \cos\gamma = \frac{z}{|\mathbf{e}|}$

3.9. Compute the scalar product of the following pairs of vectors:

a. $\mathbf{a} = [\ 0\ \ -2\ ],\ \mathbf{b} = [\ 1\ \ 3\ ]$ Answer: $\mathbf{a} \bullet \mathbf{b} = -6$

b. $\mathbf{a} = [\ 4\ \ -1\ ],\ \mathbf{b} = [\ 2\ \ 1\ ]$ Answer: $\mathbf{a} \bullet \mathbf{b} = 7$

c. $\mathbf{a} = [\ 1\ \ 0\ ],\ \mathbf{b} = [\ 0\ \ 4\ ]$ Answer: $\mathbf{a} \bullet \mathbf{b} = 0$

d. $\mathbf{a} = [\ 3\ \ 0\ \ -2\ ],\ \mathbf{b} = [\ 0\ \ -1\ \ -3\ ]$ Answer: $\mathbf{a} \bullet \mathbf{b} = 6$

e. $\mathbf{a} = [\ 5\ \ 1\ \ 7\ ],\ \mathbf{b} = [\ -2\ \ 4\ \ 1\ ]$ Answer: $\mathbf{a} \bullet \mathbf{b} = 1$

3.10. Compute the following vector products:

a. $\mathbf{i} \times \mathbf{i}$ Answer: $\mathbf{i} \times \mathbf{i} = 0$

b. $\mathbf{j} \times \mathbf{j}$ Answer: $\mathbf{j} \times \mathbf{j} = 0$

c. $\mathbf{k} \times \mathbf{k}$ Answer: $\mathbf{k} \times \mathbf{k} = 0$

d. $\mathbf{i} \times \mathbf{j}$ Answer: $\mathbf{i} \times \mathbf{j} = \mathbf{k}$

e. $\mathbf{j} \times \mathbf{k}$ Answer: $\mathbf{j} \times \mathbf{k} = \mathbf{i}$

f. $\mathbf{k} \times \mathbf{i}$ Answer: $\mathbf{k} \times \mathbf{i} = \mathbf{j}$

g. $\mathbf{j} \times \mathbf{i}$ Answer: $\mathbf{j} \times \mathbf{i} = -\mathbf{k}$

h. $\mathbf{k} \times \mathbf{j}$ Answer: $\mathbf{k} \times \mathbf{j} = -\mathbf{i}$

i. $\mathbf{i} \times \mathbf{k}$ Answer: $\mathbf{i} \times \mathbf{k} = -\mathbf{j}$

The vector product of two vectors **a** and **b** is

$$\mathbf{a} \times \mathbf{b} = (a_y b_z - a_z b_y)\mathbf{i} - (a_x b_z - a_z b_x)\mathbf{j} + (a_x b_y - a_y b_x)\mathbf{k}.$$

In matrix component form this becomes

$$\mathbf{a} \times \mathbf{b} = \begin{bmatrix} (a_y b_z - a_z b_y) & -(a_x b_z - a_z b_x) & (a_x b_y - a_y b_x) \end{bmatrix}.$$

Because **i**, **j**, and **k** are mutually perpendicular unit vectors, the computations are trivial.

3.11. Show that $\mathbf{b} \times \mathbf{a} = -(\mathbf{a} \times \mathbf{b})$.

Answer:

$$\mathbf{a} \times \mathbf{b} = (a_y b_z - a_z b_y)\mathbf{i} - (a_x b_z - a_z b_x)\mathbf{j} + (a_x b_y - a_y b_x)\mathbf{k}$$

$$\begin{aligned} -(\mathbf{a} \times \mathbf{b}) &= (a_z b_y - a_y b_z)\mathbf{i} - (a_z b_x - a_x b_z)\mathbf{j} + (a_y b_x - a_x b_y)\mathbf{k} \\ &= (\mathbf{b} \times \mathbf{a}) \end{aligned}$$

3.12. Compute the angle between the following pairs of vectors:

a. $\mathbf{a} = \begin{bmatrix} 0 & -2 \end{bmatrix}, \mathbf{b} = \begin{bmatrix} 1 & 3 \end{bmatrix}$ Answer: 161.565°

b. $\mathbf{a} = \begin{bmatrix} 4 & -1 \end{bmatrix}, \mathbf{b} = \begin{bmatrix} 2 & 1 \end{bmatrix}$ Answer: 40.601°

c. $\mathbf{a} = \begin{bmatrix} 1 & 0 \end{bmatrix}, \mathbf{b} = \begin{bmatrix} 0 & 4 \end{bmatrix}$ Answer: 90°

d. $\mathbf{a} = \begin{bmatrix} 3 & 0 & -2 \end{bmatrix}, \mathbf{b} = \begin{bmatrix} 0 & -1 & -3 \end{bmatrix}$ Answer: 58.249°

e. $\mathbf{a} = \begin{bmatrix} 5 & 1 & 7 \end{bmatrix}, \mathbf{b} = \begin{bmatrix} -2 & 4 & 1 \end{bmatrix}$ Answer: 88.556°

3.13. Compute the vector product for each of the following pairs of vectors:

a. $\mathbf{a} = \begin{bmatrix} 3 & -1 & 2 \end{bmatrix}, \mathbf{b} = \begin{bmatrix} 2 & 0 & 2 \end{bmatrix}$ Answer: $\mathbf{a} \times \mathbf{b} = \begin{bmatrix} -2 & -2 & -2 \end{bmatrix}$

b. $\mathbf{a} = \begin{bmatrix} 4 & 1 & -5 \end{bmatrix}, \mathbf{b} = \begin{bmatrix} 3 & 6 & 2 \end{bmatrix}$ Answer: $\mathbf{a} \times \mathbf{b} = \begin{bmatrix} -28 & 7 & 21 \end{bmatrix}$

c. $\mathbf{a} = \begin{bmatrix} 2 & -1 & 3 \end{bmatrix}, \mathbf{b} = \begin{bmatrix} -4 & 2 & -6 \end{bmatrix}$ Answer: $\mathbf{a} \times \mathbf{b} = \begin{bmatrix} 0 & 0 & 0 \end{bmatrix}$

d. $\mathbf{a} = \begin{bmatrix} 0 & 1 & 0 \end{bmatrix}, \mathbf{b} = \begin{bmatrix} 1 & 0 & 0 \end{bmatrix}$ Answer: $\mathbf{a} \times \mathbf{b} = \begin{bmatrix} 0 & 0 & -1 \end{bmatrix}$

e. $\mathbf{a} = \begin{bmatrix} 0 & 0 & 1 \end{bmatrix}, \mathbf{b} = \begin{bmatrix} 1 & 0 & 0 \end{bmatrix}$ Answer: $\mathbf{a} \times \mathbf{b} = \begin{bmatrix} 0 & 1 & 0 \end{bmatrix}$

3.14. Show why the vectors $\mathbf{a} = \begin{bmatrix} -\frac{1}{3} & \frac{2}{3} & \frac{2}{3} \end{bmatrix}$, $\mathbf{b} = \begin{bmatrix} \frac{2}{3} & -\frac{1}{3} & \frac{2}{3} \end{bmatrix}$, and $\mathbf{c} = \begin{bmatrix} -\frac{2}{3} & -\frac{2}{3} & \frac{1}{3} \end{bmatrix}$ are mutually perpendicular.

Answer: They are mutually perpendicular because $\mathbf{a} \bullet \mathbf{b} = 0$, $\mathbf{a} \bullet \mathbf{c} = 0$, and $\mathbf{b} \bullet \mathbf{c} = 0$.

3.15. Show that the line joining the midpoints of two sides of a triangle is parallel to the third side and has one-half its magnitude.

Answer: Let **a** and **b** be two sides of a triangle; then **b** – **a** is the third side. The midpoints of **a** and **b** are 0.5**a** and 0.5**b**, respectively. A line joining these midpoints is represented by the vector 0.5 (**b** – **a**), which is parallel to and half the length of (**b** – **a**).

3.16. Is the set of all vectors lying in the first quadrant of the *xy* plane a linear space? Why?

Answer: The set does not define a linear vector space because it is not closed with respect to multiplication by real numbers k. If $k < 0$, then k_p does not belong to the first quadrant.

3.17. Determine nontrivial linear relations for the following sets of vectors:

a. $\mathbf{p} = [\ 1 \quad 0 \quad -2\ ],\ \mathbf{q} = [\ 3 \quad -1 \quad 3\ ],\ \mathbf{r} = [\ 5 \quad -2 \quad 8\ ]$

Answer: $\mathbf{p} - 2\mathbf{q} + \mathbf{r} = 0$

b. $\mathbf{p} = [\ 2 \quad 0 \quad 1\ ],\ \mathbf{q} = [\ 0 \quad 5 \quad 1\ ],\ \mathbf{r} = [\ 6 \quad -5 \quad 4\ ]$

Answer: $3\mathbf{p} + \mathbf{q} - \mathbf{r} = 0$

c. $\mathbf{p} = [\ 3 \quad 0\ ],\ \mathbf{q} = [\ 1 \quad 4\ ],\ \mathbf{r} = [\ 2 \quad -1\ ]$

Answer: $3\mathbf{p} - \mathbf{q} + 4\mathbf{r} = 0$

3.18. Are the vectors $\mathbf{r} = [\ 1 \quad -1 \quad 0\ ]$, $\mathbf{s} = [\ 0 \quad 2 \quad -1\ ]$, and $\mathbf{t} = [\ 2 \quad 0 \quad -1\ ]$ linearly dependent? Why?

Answer: They are linearly dependent, because 2**r** + **s** – **t** = 0.

A set of vectors $\mathbf{x}_1, \mathbf{x}_2, \ldots, \mathbf{x}_n$ is linearly dependent if and only if there are real numbers $a_1, a_2, \ldots, a_n$ not all equal to zero, such that

$$a_1\mathbf{x}_1 + a_2\mathbf{x}_2 + \ldots + a_n\mathbf{x}_n = 0.$$

If this equation is true only if $a_1, a_2, \ldots, a_n$ are all zero, then $\mathbf{x}_1, \mathbf{x}_2, \ldots, \mathbf{x}_n$ are linearly independent.

3.19. Take the vector product of two vectors **a** and **b** in two dimensions. The resulting vector has a magnitude $ab\sin\theta$. What about its direction? Where does it point? Remember, this is a two-dimensional problem.

Answer: The vector product of two vectors **p** and **r** in a two-dimensional space has a magnitude $|\mathbf{p}||\mathbf{r}|\sin\theta$, but it points nowhere. There is no third dimension, mutually perpendicular to **p** and **r**, in which it can point. The result is not a vector.

3.20. Repeat Exercise 3.19 for the vector product in four dimensions.

Answer: The vector product of two vectors **p** and **r** in a four-dimensional space appears to be a plane, because this is the locus of an infinite number of lines perpendicular to both **p** and **r**. The assignment of a specific direction to the vector product in three-dimensional space turns out to be a special case.

3.21. Show that $(\mathbf{p}-\mathbf{q})\bullet(\mathbf{p}+\mathbf{q})=|\mathbf{p}|^2-|\mathbf{q}|^2$ and that $(\mathbf{p}-\mathbf{q})\times(\mathbf{p}+\mathbf{q})=2\mathbf{p}\times\mathbf{q}$. Interpret the results with an appropriate sketch.

Answer:

$$
\begin{aligned}
(\mathbf{p}-\mathbf{q})\bullet(\mathbf{p}+\mathbf{q}) &= \mathbf{p}\bullet(\mathbf{p}+\mathbf{q})-\mathbf{q}\bullet(\mathbf{p}+\mathbf{q})\\
&= \mathbf{p}\bullet\mathbf{p}+\mathbf{p}\bullet\mathbf{q}-\mathbf{q}\bullet\mathbf{p}-\mathbf{q}\bullet\mathbf{q}\\
&= \mathbf{p}\bullet\mathbf{p}-\mathbf{q}\bullet\mathbf{q}\\
&= |\mathbf{p}|^2-|\mathbf{q}|^2
\end{aligned}
$$

and

$$
\begin{aligned}
(\mathbf{p}-\mathbf{q})\times(\mathbf{p}+\mathbf{q}) &= (\mathbf{p}-\mathbf{q})\times\mathbf{p}+(\mathbf{p}-\mathbf{q})\times\mathbf{q}\\
&= \mathbf{p}\times\mathbf{p}-\mathbf{q}\times\mathbf{p}+\mathbf{p}\times\mathbf{q}-\mathbf{q}\times\mathbf{p}\\
&= -2\mathbf{q}\times\mathbf{p}\\
&= 2\mathbf{p}\times\mathbf{q}
\end{aligned}
$$

3.22. Given vectors $\mathbf{a}=6\mathbf{i}+10\mathbf{j}+2\mathbf{k}$ and $\mathbf{b}=\mathbf{i}+2\mathbf{j}+6\mathbf{k}$, find $\mathbf{c}=\mathbf{a}+\mathbf{b}$ (**i**, **j**, and **k** are unit vectors in the x, y, and z directions, respectively).

Answer: $\mathbf{c} = 7\mathbf{i} + 8\mathbf{j} + 8\mathbf{k}$

3.23. Given that **p** and **q** are linearly independent vectors in the plane, find the value of k that makes each of the following pairs of vectors collinear:

a. $k\mathbf{p} + 2\mathbf{q}, \mathbf{p} - \mathbf{q}$ Answer: $k = -2$

b. $(k + 1)\mathbf{p} + \mathbf{q}, 2\mathbf{q}$ Answer: $k = -1$

c. $k\mathbf{p} + \mathbf{q}, \mathbf{p} + k\mathbf{q}$ Answer: $k = 1$

3.24. Show that the magnitude of a unit vector is equal to 1.

Answer: If $\hat{\mathbf{a}} = \mathbf{a}/|\mathbf{a}|$, then $|\hat{\mathbf{a}}| = \sqrt{\left(\frac{a_x}{|\mathbf{a}|}\right)^2+\left(\frac{a_y}{|\mathbf{a}|}\right)^2+\left(\frac{a_z}{|\mathbf{a}|}\right)^2}$,

or

$$|\hat{\mathbf{a}}| = \sqrt{\frac{a_x^2}{|\mathbf{a}|^2}+\frac{a_y^2}{|\mathbf{a}|^2}+\frac{a_z^2}{|\mathbf{a}|^2}} = \frac{\sqrt{a_x^2+a_y^2+a_z^2}}{|\mathbf{a}|} = 1.$$

3.25. Find the magnitude and direction numbers for each of the following vectors:

a. $\mathbf{a} = (3, 4)$ Answer: $|\mathbf{a}| = 5$; 0.6, 0.8

b. $\mathbf{b} = (0, -2)$ Answer: $|\mathbf{b}| = 2$; 0, −1

c. $\mathbf{c} = (-3, -5, 0)$ Answer: $|\mathbf{c}| = \sqrt{34}$; $-\frac{3}{\sqrt{34}}, -\frac{5}{\sqrt{34}}, 0$

d. $\mathbf{d} = (1, 4, -3)$ Answer: $|\mathbf{d}| = \sqrt{26}$; $\frac{1}{\sqrt{26}}, \frac{4}{\sqrt{26}}, -\frac{3}{\sqrt{26}}$

e. $\mathbf{e} = (x, y, z)$ Answer: $|\mathbf{e}| = \sqrt{x^2 + y^2 + z^2}$; $\frac{x}{|\mathbf{e}|}, \frac{y}{|\mathbf{e}|}, \frac{z}{|\mathbf{e}|}$

3.26. Find the inner product of the following pairs of vectors:

a. $[\ 0\ \ -2\ ], [\ 1\ \ 3\ ]$ Answer: –6

b. $[\ 4\ \ -1\ ], [\ 2\ \ 1\ ]$ Answer: 7

c. $[\ 1\ \ 0\ ], [\ 0\ \ 4\ ]$ Answer: 0

d. $[\ 3\ \ 0\ \ -2\ ], [\ 0\ \ -1\ \ -3\ ]$ Answer: 6

e. $[\ 5\ \ 1\ \ 7\ ], [\ -2\ \ 4\ \ 1\ ]$ Answer: 1

3.27. Find the angle between each pair of vectors given in Exercise 3.27.

Answer:

a. 18.435°

b. 40.601°

c. 90°

d. 58.249°

e. 88.556°

3.28. Find the vector product for each of the following pairs of vectors:

a. $[\ 3\ \ -1\ \ 2\ ], [\ 2\ \ 0\ \ 2\ ]$ Answer: $[\ -2\ \ -2\ \ -2\ ]$

b. $[\ 4\ \ 1\ \ -5\ ], [\ 3\ \ 6\ \ 2\ ]$ Answer: $[\ -28\ \ 7\ \ 21\ ]$

c. $[\ 2\ \ -1\ \ 3\ ], [\ -4\ \ 2\ \ -6\ ]$ Answer: $[\ 0\ \ 0\ \ 0\ ]$

d. $[\ 0\ \ 1\ \ 0\ ], [\ 1\ \ 0\ \ 0\ ]$ Answer: $[\ 0\ \ 0\ \ -1\ ]$

e. $[\ 0\ \ 0\ \ 1\ ], [\ 1\ \ 0\ \ 0\ ]$ Answer: $[\ 0\ \ 1\ \ 0\ ]$

Chapter 4 Matrix Basics

4.1. Given $\mathbf{A} = \begin{bmatrix} 7 & 3 & -1 \\ 2 & -5 & 6 \end{bmatrix}$ and $\mathbf{B} = \begin{bmatrix} 1 & 5 & 6 \\ -4 & -2 & 3 \end{bmatrix}$,

a. Find **A** + **B**. Answer: $\mathbf{A}+\mathbf{B}=\begin{bmatrix} 8 & 8 & 5 \\ -2 & -7 & 9 \end{bmatrix}$

b. Find **A** – **B**. Answer: $\mathbf{A}-\mathbf{B}=\begin{bmatrix} 6 & -2 & -7 \\ 6 & -3 & 3 \end{bmatrix}$

Adding two matrices **A** and **B** produces a third matrix **C**, whose elements are equal to the sum of the corresponding elements of **A** and **B**:

$a_{ij} + b_{ij} = c_{ij}$.

4.2. Given $\mathbf{A}=[\ 3 \quad 7 \quad -2\ ]$, find –**A**.

Answer: $-\mathbf{A}=[\ -3 \quad -7 \quad 2\ ]$

4.3. Given $\mathbf{A}=\begin{bmatrix} 1 & 5 & 2 \\ 0 & -1 & 4 \end{bmatrix}$, $\mathbf{B}=\begin{bmatrix} 6 & 1 & 3 \\ 0 & 9 & 2 \end{bmatrix}$, and $\mathbf{C}=\begin{bmatrix} 4 & 1 & 1 \\ 5 & 8 & 3 \end{bmatrix}$,

a. Find **A** + 2**A**. Answer: $\mathbf{A}+2\mathbf{A}=3\mathbf{A}=\begin{bmatrix} 3 & 15 & 6 \\ 0 & -3 & 12 \end{bmatrix}$

b. Find **B** + **B**. Answer: $\mathbf{B}+\mathbf{B}=2\mathbf{B}=\begin{bmatrix} 12 & 2 & 6 \\ 0 & 18 & 4 \end{bmatrix}$

c. Find 2**A** + **B**. Answer: $2\mathbf{A}+\mathbf{B}=\begin{bmatrix} 8 & 11 & 7 \\ 0 & 7 & 10 \end{bmatrix}$

d. Find **A** – **B** + **C**. Answer: $\mathbf{A}-\mathbf{B}+\mathbf{C}=\begin{bmatrix} -1 & 5 & 0 \\ 5 & 5 & 5 \end{bmatrix}$

e. Find **A** – 2**B** – **C**. Answer: $\mathbf{A}-2\mathbf{B}-\mathbf{C}=\begin{bmatrix} -15 & 2 & -5 \\ -5 & -27 & -3 \end{bmatrix}$

4.4. Given $\mathbf{A}=\begin{bmatrix} 1 & -4 \\ 3 & 0 \end{bmatrix}$, find 1.5**A**. Answer: $1.5\mathbf{A}=\begin{bmatrix} 1.5 & -6 \\ 4.5 & 0 \end{bmatrix}$

4.5. Find $\mathbf{I}^T$. Answer: $\mathbf{I}^T = \mathbf{I}$

As presented earlier in this chapter, we see that the identity matrix is very important in matrix algebra. It is a special square and diagonal matrix that has unit elements on the main diagonal. The symbol **I** denotes it. The 3×3 identity matrix is

$$\mathbf{I}=\begin{bmatrix} 1 & 0 & 0 \\ 0 & 1 & 0 \\ 0 & 0 & 1 \end{bmatrix}.$$

Recall that the Greek symbol with subscripts, δ_{ij}, denotes the elements of **I**, where

$$\delta_{ij} = 0 \text{ if } i \neq j,$$

$$\delta_{ij} = 1 \text{ if } i = j.$$

By interchanging the rows and columns of a matrix **A**, we obtain its transpose, $\mathbf{A}^{\mathrm{T}}$, so that $a_{ij}^{\mathrm{T}} = a_{ji}$, where the a_{ij}^{T} are the elements of the transpose of **A**. For example, if

$$\mathbf{A} = \begin{bmatrix} a & c & e \\ b & d & f \end{bmatrix}, \text{ then } \mathbf{A}^{\mathrm{T}} = \begin{bmatrix} a & b \\ c & d \\ e & f \end{bmatrix}.$$

4.6. Given $\mathbf{A} = \begin{bmatrix} 5 & 3 & 8 \\ -1 & 4 & 7 \\ 0 & 1 & 1 \end{bmatrix}$ and $\mathbf{B} = \begin{bmatrix} 6 & 7 \\ 10 & 9 \\ 2 & -3 \end{bmatrix}$, find **AB**.

Answer: $\mathbf{AB} = \begin{bmatrix} 76 & 38 \\ 48 & 8 \\ 12 & 6 \end{bmatrix}$

Remember that the product **AB** of two matrices is another matrix **C**. This operation is possible if and only if the number of columns of the first matrix is equal to the number of rows of the second matrix. It **A** is $m \times n$ and **B** is $n \times p$, then **C** is $m \times p$. When this condition is satisfied, we say that the matrices are conformable for multiplication.

4.7. Given $\mathbf{A} = \begin{bmatrix} 2 & 1 \\ 3 & 4 \end{bmatrix}$ and $\mathbf{B} = \begin{bmatrix} 6 \\ 3 \end{bmatrix}$, find **AB**. Answer: $\mathbf{AB} = \begin{bmatrix} 15 \\ 30 \end{bmatrix}$

4.8. Given $\mathbf{A} = \begin{bmatrix} 4 & 0 & 7 \\ 5 & 1 & 2 \end{bmatrix}$, find $\mathbf{A}^{\mathrm{T}}$. Answer: $\mathbf{A}^{\mathrm{T}} = \begin{bmatrix} 4 & 5 \\ 0 & 1 \\ 7 & 2 \end{bmatrix}$

4.9. Given $\mathbf{A} = \begin{bmatrix} 4 & -2 \\ 1 & 0 \\ 6 & 7 \end{bmatrix}$ and $\mathbf{B} = \begin{bmatrix} 1 & 1 \\ 5 & 2 \\ 2 & 4 \end{bmatrix}$,

a. Find $\left(\mathbf{A}^{\mathrm{T}}\right)^{\mathrm{T}}$. Answer: $\left(\mathbf{A}^{\mathrm{T}}\right)^{\mathrm{T}} = \mathbf{A}$

b. Find $(\mathbf{A}+\mathbf{B})^{\mathrm{T}}$. Answer: $(\mathbf{A}+\mathbf{B})^{\mathrm{T}} = \begin{bmatrix} 5 & 6 & 8 \\ -1 & 2 & 11 \end{bmatrix}$

c. Find $\mathbf{A}^{\mathrm{T}} + \mathbf{B}^{\mathrm{T}}$. Answer: $\mathbf{A}^{\mathrm{T}} + \mathbf{B}^{\mathrm{T}} = (\mathbf{A}+\mathbf{B})^{\mathrm{T}} = \begin{bmatrix} 5 & 6 & 8 \\ -1 & 2 & 11 \end{bmatrix}$

d. Find $\mathbf{B}^{\mathrm{T}} + \mathbf{A}^{\mathrm{T}}$. Answer: $\mathbf{B}^{\mathrm{T}} + \mathbf{A}^{\mathrm{T}} = \mathbf{A}^{\mathrm{T}} + \mathbf{B}^{\mathrm{T}}$

4.10. Find the product $\begin{bmatrix} t^2 & t & 1 \end{bmatrix} \begin{bmatrix} a_x & a_y \\ b_x & b_y \\ c_x & c_y \end{bmatrix}$.

Answer: $\begin{bmatrix} t^2 & t & 1 \end{bmatrix} \begin{bmatrix} a_x & a_y \\ b_x & b_y \\ c_x & c_y \end{bmatrix} = \left[(a_x t^2 + b_x t + c_x)(a_y t^2 + b_y t + c_y) \right]$

4.11. Given $\mathbf{A} = \begin{bmatrix} 1 & 0 & 0 \\ 0 & 1 & 0 \\ 0 & 0 & 1 \end{bmatrix}$, $\mathbf{B} = \begin{bmatrix} 7 \\ 4 \\ 9 \\ 5 \end{bmatrix}$, and $\mathbf{C} = [\ 1 \quad -2 \quad 4 \quad 6\ ]$,

a. Find a_{23}. Answer: $a_{23} = 0$
b. Find a_{32}. Answer: $a_{32} = 0$
c. Find b_{31}. Answer: $b_{31} = 9$
d. Find c_{14}. Answer: $c_{14} = 6$
e. What is the order of **A**? Answer: 3×3
f. What is the order of **B**? Answer: 4×1
g. What is the order of **C**? Answer: 1×4
h. Which is the column matrix? Answer: **B**
i. Which is the row matrix? Answer: **C**
j. Which is the identity matrix? Answer: **A** = **I**

4.12. If **P** = **ABC** and the order of **A** is 1×4, the order of **B** is 4×4, and the order of **C** is 4×3, then what is the order of **P**?
Answer: **P** is 1×3.

Chapter 5 Special Matrices

5.1. Given $\mathbf{A} = \begin{bmatrix} 7 & 4 & 4 \\ 9 & 1 & 3 \\ 0 & 2 & 5 \end{bmatrix}$ and $\mathbf{B} = \begin{bmatrix} 6 & 5 \\ 8 & 1 \\ 3 & 9 \end{bmatrix}$,

a. Find a_{23}. Answer: 3
b. Find a_{12}. Answer: 4
c. Find a_{31}. Answer: 0
d. Find b_{11}. Answer: 6
e. Find b_{32}. Answer: 9

f. What is the order of **A**? Answer: 3×3

g. What is the order of **B**? Answer: 3×2

h. Which matrix, if any, is a square matrix? Answer: **A**

i. List the elements, in order, on the main diagonal of **A**.

Answer: $a_{11} = 7, a_{22} = 1, a_{33} = 5$

j. Change a_{12}, a_{13} and a_{23} so that **A** is a symmetric matrix.

Answer: $a_{12} = 9, a_{13} = 0, a_{23} = 2$

5.2. Find the values of the following δ_{ij}:

a. δ_{32} Answer: $\delta_{32} = 0$

b. δ_{14} Answer: $\delta_{14} = 0$

c. δ_{33} Answer: $\delta_{33} = 1$

d. $\delta_{7,10}$ Answer: $\delta_{7,10} = 0$

e. δ_{11} Answer: $\delta_{11} = 1$

A very important matrix in matrix algebra is the identity matrix. This is a special diagonal matrix that has unit elements on the main diagonal. It is denoted by the symbol **I**. The 3×3 identity matrix is

$$\mathbf{I} = \begin{bmatrix} 1 & 0 & 0 \\ 0 & 1 & 0 \\ 0 & 0 & 1 \end{bmatrix}.$$

Elements of **I** are denoted by δ_{ij}, where

$$\begin{aligned} \delta_{ij} &= 0 \text{ if } i \neq j \\ &= 1 \text{ if } i = j. \end{aligned}$$

5.3. Given $\mathbf{A} = \begin{bmatrix} 5 & 4 & 9 \\ 2 & 1 & 0 \\ 6 & 7 & 1 \end{bmatrix}$,

a. Change a_{12}, a_{13}, and a_{23} so that **A** becomes antisymmetric.

Answer: $a_{12} = -2,\ a_{13} = -6,\ a_{23} = -7$

b. What other changes, if any, are necessary?

Answer: $a_{11} = a_{22} = a_{33} = 0$

A matrix whose elements are symmetric about the main diagonal is a symmetric matrix. If **A** is a symmetric matrix, then $a_{ij} = a_{ji}$.

5.4. Write out the 2×2 null matrix.

Answer: $\begin{bmatrix} 0 & 0 \\ 0 & 0 \end{bmatrix}$

5.5. Are the following matrices orthogonal, proper, or improper?

a. $\begin{bmatrix} \frac{\sqrt{2}}{2} & \frac{\sqrt{2}}{2} \\ -\frac{\sqrt{2}}{2} & \frac{\sqrt{2}}{2} \end{bmatrix}$

Answer: $\begin{vmatrix} \frac{\sqrt{2}}{2} & \frac{\sqrt{2}}{2} \\ -\frac{\sqrt{2}}{2} & \frac{\sqrt{2}}{2} \end{vmatrix} = +1$. Proper and orthogonal.

b. $\begin{bmatrix} 1 & \frac{1}{2} \\ 2 & 0 \end{bmatrix}$

Answer: $\begin{vmatrix} 1 & \frac{1}{2} \\ 2 & 0 \end{vmatrix} = -1$. Improper and orthogonal.

c. $\begin{bmatrix} 3 & 1 \\ 5 & 2 \end{bmatrix}$

Answer: $\begin{vmatrix} 3 & 1 \\ 5 & 2 \end{vmatrix} = +1$. Proper and orthogonal.

d. $\begin{bmatrix} -\frac{\sqrt{3}}{2} & \frac{1}{2} \\ \frac{1}{2} & \frac{\sqrt{3}}{2} \end{bmatrix}$

Answer: $\begin{vmatrix} -\frac{\sqrt{3}}{2} & \frac{1}{2} \\ \frac{1}{2} & \frac{\sqrt{3}}{2} \end{vmatrix} = -1$. Improper and orthogonal.

e. $\begin{bmatrix} 0 & 0 & 1 \\ 1 & 0 & 0 \\ 0 & -1 & 0 \end{bmatrix}$

Answer: $\begin{vmatrix} 0 & 0 & 1 \\ 1 & 0 & 0 \\ 0 & -1 & 0 \end{vmatrix} = -1$. Improper and orthogonal.

f. $\begin{bmatrix} \frac{1}{2} & -\frac{\sqrt{3}}{2} & 0 \\ \frac{\sqrt{3}}{2} & \frac{1}{2} & 0 \\ 0 & 0 & 2 \end{bmatrix}$

Answer: $\begin{vmatrix} \frac{1}{2} & -\frac{\sqrt{3}}{2} & 0 \\ \frac{\sqrt{3}}{2} & \frac{1}{2} & 0 \\ 0 & 0 & 2 \end{vmatrix} = +1$. Proper and orthogonal.

g. $\begin{bmatrix} \frac{\sqrt{3}}{2} & \frac{\sqrt{3}}{4} & 1 \\ \frac{1}{2} & -\frac{3}{4} & -\frac{\sqrt{3}}{4} \\ 0 & \frac{1}{2} & -\frac{\sqrt{3}}{2} \end{bmatrix}$

Answer: $\begin{vmatrix} \frac{\sqrt{3}}{2} & \frac{\sqrt{3}}{4} & 1 \\ \frac{1}{2} & -\frac{3}{4} & -\frac{\sqrt{3}}{4} \\ 0 & \frac{1}{2} & -\frac{\sqrt{3}}{2} \end{vmatrix} = +1$. Proper and orthogonal.

5.6. Find $\begin{bmatrix} 0 & 0 & 1 \\ 1 & 0 & 0 \\ 0 & -1 & 0 \end{bmatrix}^{-1}$.

Answer: $\begin{bmatrix} 0 & 0 & 1 \\ 1 & 0 & 0 \\ 0 & -1 & 0 \end{bmatrix}^{-1} = \begin{bmatrix} 0 & 1 & 0 \\ 0 & 0 & -1 \\ 1 & 0 & 0 \end{bmatrix}$

Matrix arithmetic does not define a division operation, but it does include a process for finding the inverse of a matrix. The inverse of a square matrix **A** is $\mathbf{A}^{-1}$, which satisfies the condition

$$\mathbf{A}\mathbf{A}^{-1} = \mathbf{A}^{-1}\mathbf{A} = \mathbf{I}.$$

The elements of $\mathbf{A}^{-1}$ are a_{ij}^{-1}, where

$$a_{ij}^{-1} = \frac{(-1)^{i+j} \,|\, \mathbf{A}'_{ji} \,|}{|\, \mathbf{A} \,|}$$

and where the prime symbol denotes a minor of the determinant.

5.7. Show that the inverse of the orthogonal matrix **A** is an orthogonal matrix, where

$$\mathbf{A} = \begin{bmatrix} \frac{2}{3} & -\frac{2}{3} & \frac{1}{3} \\ \frac{1}{3} & \frac{2}{3} & \frac{2}{3} \\ \frac{2}{3} & \frac{1}{3} & -\frac{2}{3} \end{bmatrix}. \qquad \text{Answer: } \left|\mathbf{A}^{-1}\right| = -1.$$

5.8. Compute the following determinants:

a. $\begin{vmatrix} 2 & 0 \\ -3 & 2 \end{vmatrix}$ Answer: $\begin{vmatrix} 2 & 0 \\ -3 & 2 \end{vmatrix} = 4$

b. $\begin{vmatrix} 1 & 2 \\ 4 & -5 \end{vmatrix}$ Answer: $\begin{vmatrix} 1 & 2 \\ 4 & -5 \end{vmatrix} = -13$

c. $\begin{vmatrix} 0 & 3 \\ 0 & 0 \end{vmatrix}$ Answer: $\begin{vmatrix} 0 & 3 \\ 0 & 0 \end{vmatrix} = 0$

d. $\begin{vmatrix} 1 & 2 \\ 2 & 4 \end{vmatrix}$ Answer: $\begin{vmatrix} 1 & 2 \\ 2 & 4 \end{vmatrix} = 0$

e. $\begin{vmatrix} 2 & 5 \\ -3 & 1 \end{vmatrix}$ Answer: $\begin{vmatrix} 2 & 5 \\ -3 & 1 \end{vmatrix} = 17$

Recall that the determinant of a 2×2 matrix **A** is written as $|\mathbf{A}|$ and

$|\mathbf{A}| = a_{11}a_{22} - a_{12}a_{21}$.

5.9. Given $|\mathbf{A}| = \begin{bmatrix} 4 & 0 & -1 \\ 1 & 2 & 1 \\ -3 & 6 & 5 \end{bmatrix}$, compute the following minors and cofactors:

a. m_{11} Answer: $m_{11} = 4$

b. m_{21} Answer: $m_{21} = 6$

c. m_{31} Answer: $m_{31} = 2$

d. m_{22} Answer: $m_{22} = 17$

e. m_{12} Answer: $m_{12} = 8$

f. c_{11} Answer: $c_{11} = 4$

g. c_{21} Answer: $c_{21} = -6$

h. c_{31} Answer: $c_{31} = 2$

i. c_{22} Answer: $c_{22} = 17$

j. c_{12} Answer: $c_{12} = -8$

5.10. Compute $|\mathbf{A}|$ in the preceding exercise.

Answer: $|\mathbf{A}| = 4$

5.11. Compute the inverse of the following matrices, if one exists:

a. $\begin{bmatrix} 1 & 0 \\ 0 & 1 \end{bmatrix}$ Answer: $\begin{bmatrix} 1 & 0 \\ 0 & 1 \end{bmatrix}^{-1} = \begin{bmatrix} 1 & 0 \\ 0 & 1 \end{bmatrix}$

b. $\begin{bmatrix} 3 & -1 & 2 \\ 1 & 2 & 1 \\ -2 & 1 & 3 \end{bmatrix}$ Answer: $\begin{bmatrix} 3 & -1 & 2 \\ 1 & 2 & 1 \\ -2 & 1 & 3 \end{bmatrix}^{-1} = \frac{1}{30}\begin{bmatrix} 5 & 5 & -5 \\ -5 & 13 & -1 \\ 5 & -1 & 7 \end{bmatrix}$

c. $\begin{bmatrix} 1 & 0 & 0 \\ 2 & 1 & 3 \\ 1 & 1 & 2 \end{bmatrix}$ Answer: $\begin{bmatrix} 1 & 0 & 0 \\ 2 & 1 & 3 \\ 1 & 1 & 2 \end{bmatrix}^{-1} = \begin{bmatrix} 1 & 0 & 0 \\ 1 & -2 & 3 \\ -1 & 1 & -1 \end{bmatrix}$

d. $\begin{bmatrix} 3 & -1 & 2 \\ 1 & 2 & 1 \\ 3 & -1 & 2 \end{bmatrix}$

Answer: The inverse does not exist because $\begin{vmatrix} 3 & -1 & 2 \\ 1 & 2 & 1 \\ 3 & -1 & 2 \end{vmatrix} = 0.$

5.12. Find the eigenvalues and eigenvectors of the following matrices:

a. $\begin{bmatrix} 3 & 5 \\ 4 & 5 \end{bmatrix}$

Answer: The eigenvalues are 7 and –5, corresponding to the eigenvectors $[\ k \quad 2k\ ]^T$ and $[\ 5k \quad -2k\ ]^T$, respectively.

b. $\begin{bmatrix} 1 & 2 \\ -2 & 5 \end{bmatrix}$

Answer: The eigenvalue is 3, corresponding to the eigenvector $[\ k \quad k\ ]^T$.

5.13. Find the characteristic equation, eigenvalues, and corresponding eigenvectors for the following matrices:

a. $\begin{bmatrix} 1 & 2 \\ 4 & 3 \end{bmatrix}$

Answer: The characteristic equation is $\lambda^2 - 4\lambda - 5 = 0$, and the eigenvalues are 5 and –1, corresponding to the eigenvectors $[\ k \quad 2k\ ]^T$ and $[\ k \quad -k\ ]^T$, respectively.

b. $\begin{bmatrix} 2 & 0 & 0 \\ 0 & 1 & 0 \\ 0 & 0 & 3 \end{bmatrix}$

Answer: The characteristic equation is $\lambda^3 - 6\lambda^2 + 11\lambda - 6 = 0$, and the eigenvalues are 2, 1, and 3, corresponding to the eigenvectors $[\ k \quad 0 \quad 0\ ]^T$, $[\ 0 \quad k \quad 0\ ]^T$, and $[\ 0 \quad 0 \quad k\ ]^T$, respectively.

5.14. Prove that $|\mathbf{AB}| = |\mathbf{BA}| = |\mathbf{A}||\mathbf{B}| = |\mathbf{B}||\mathbf{A}|$ for any two square matrices **A** and **B**. (Note that this is true for square matrices of any order.)

Answer: Let $\mathbf{A} = \begin{bmatrix} a & b \\ c & d \end{bmatrix}$ and $\mathbf{B} = \begin{bmatrix} e & f \\ g & h \end{bmatrix}$. Then

$\mathbf{AB} = \begin{bmatrix} ae+bg & af+bh \\ ce+dg & cf+dh \end{bmatrix}$ and

$\mathbf{BA} = \begin{bmatrix} ae+cf & be+df \\ ag+ch & bg+dh \end{bmatrix}$, and also with $|\mathbf{A}| = ad - bc$, $|\mathbf{B}| = ef + fg$, and

$|\mathbf{A}||\mathbf{B}| = adeh - adfg - bceh + bcfg$, $|\mathbf{AB}| = adeh - adfg - bceh + bcfg$, etc.

Chapter 8 Translation and Rotation

8.1. Find appropriate translation pairs Δx and Δy such that the algebraic equation of a straight line in slope-intercept form is simplified (that is, the line is translated so that it passes through the origin).

Answer: In slope-intercept form, the equation describing the translation of a straight line is $y' = mx' + (b - m\Delta x + \Delta y)$, where m and b are the slope and intercept, respectively, of the original line. For a line passing through the origin, it must be true that $y' = mx'$. This means that the intercept must equal zero (i.e., $b - m\Delta x + \Delta y = 0$). One way this can happen is if $\Delta x = 0$, $\Delta y = b$; another is if $\Delta x = b/m$, $\Delta y = 0$.

8.2. Using vectors, show that a translated line is parallel to its original position.

Answer: Translate both the original and its translated copy so that they both pass through the origin, producing $\mathbf{r}' = u\mathbf{t}$ and $\mathbf{r}' = u\mathbf{t}$.

8.3. How far is each point moved when translated $x' = x + a,\ y' = y + b,\ z' = z + c$?
Answer: $d = +\sqrt{a^2 + b^2 + c^2}$

8.4. Find the equations of the translation that sends (–4, 3) into (1, 2).
Answer: $x' = x + 5$ and $y' = y - 1$.

8.5. Find the transformed equations of the following equations subject to the transformation $x' = x + 2,\ y' = y - 1$.

a. $x + y - 1 = 0$ Answer: $x' + y' - 2 = 0$

b. $x^2 + y^2 = 4$ Answer: $x'^2 + y'^2 - 4x' + 2y' = -1$

c. $y = 2x^2$ Answer: $y' = 2x'^2 - 8x' + 7$

8.6. Express each of the following translations in vector form, giving the components of each vector:

a. $x' = x + 5,\ y' = y$ Answer: **t** = (5, 0)

b. $x' = x - 3,\ y' = y + 2$ Answer: **t** = (–3, 2)

c. $x' = x,\ y' = y,\ z' = z$ Answer: **t** = (0, 0, 0)

d. $x' = x - 1,\ y' = y,\ z' = z + 6$ Answer: **t** = (–1, 0, 6)

e. $x' = x + a,\ y' = y + b,\ z' = z + c$ Answer: **t** = (a, b, c)

8.7. Find the rotation matrix for rotations of (a) 30°, (b) 45°, (c) 90°, and (d) 180° in the xy plane about the origin.
Answer:

a. $\mathbf{R}_{30} = \begin{bmatrix} 0.866 & -0.500 \\ 0.500 & 0.866 \end{bmatrix}$

b. $\mathbf{R}_{45} = \begin{bmatrix} 0.707 & -0.707 \\ 0.707 & 0.707 \end{bmatrix}$

c. $\mathbf{R}_{90} = \begin{bmatrix} 0 & -1 \\ 1 & 0 \end{bmatrix}$

d. $\mathbf{R}_{180} = \begin{bmatrix} -1 & 0 \\ 0 & -1 \end{bmatrix}$

8.8. Compute the determinant of each of the rotation matrices found for the exercise above.
Answer:

a. $|\mathbf{R}_{30}| = 1$

b. $|\mathbf{R}_{45}| = 1$

c. $|\mathbf{R}_{90}| = 1$

d. $|\mathbf{R}_{180}| = 1$

8.9. Find the algebraic equations for rotating a point x, y through an angle θ about x_c, y_c.

Answer:

$$x' = (x - x_c)\cos\theta - (y - y_c)\sin\theta + x_c$$

and

$$y' = (x - x_c)\sin\theta - (y - y_c)\cos\theta + y_c$$

8.10. Consider successive rotations about two different points. Is the outcome independent of the order in which we perform the rotations? Explain your answer.

Answer:

Rotating θ_1 about $\mathbf{p}_{c1}$, followed by θ_2 about $\mathbf{p}_{c2}$, produces

$$\mathbf{p}''_{12} = \mathbf{R}_{\theta_2}\left[\mathbf{R}_{\theta_1}(\mathbf{p} - \mathbf{p}_{c1}) + \mathbf{p}_{c1} - \mathbf{p}_{c2}\right] + \mathbf{p}_{c2}.$$

Rotating θ_2 about $\mathbf{p}_{c2}$, followed by θ_1 about $\mathbf{p}_{c1}$, produces

$$\mathbf{p}''_{21} = \mathbf{R}_{\theta_1}\left[\mathbf{R}_{\theta_2}(\mathbf{p} - \mathbf{p}_{c2}) + \mathbf{p}_{c2} - \mathbf{p}_{c1}\right] + \mathbf{p}_{c1}.$$

In general, $\mathbf{p}''_{12} \neq \mathbf{p}''_{21}$, so that order is important.

8.11. Verify

$$\mathbf{R}_{\varphi\theta\phi} = \begin{bmatrix} \cos\theta\cos\phi & -\cos\theta\sin\phi & \sin\theta \\ \cos\varphi\sin\phi + \sin\varphi\sin\theta\cos\phi & \cos\varphi\cos\phi - \sin\varphi\sin\theta\sin\phi & -\sin\varphi\cos\theta \\ \sin\varphi\sin\phi - \cos\varphi\sin\theta\cos\phi & \sin\varphi\cos\phi + \cos\varphi\sin\theta\sin\phi & \cos\varphi\cos\theta \end{bmatrix}$$

by setting

a. $\alpha = \beta = 0$

Answer: $\alpha = \beta = 0$ and $\mathbf{R}_{zyx} = \mathbf{R}_z = \begin{bmatrix} \cos\gamma & -\sin\gamma & 0 \\ \sin\gamma & \cos\gamma & 0 \\ 0 & 0 & 1 \end{bmatrix}$

b. $\alpha = \gamma = 0$

Answer: $\alpha = \gamma = 0$ and $\mathbf{R}_{zyx} = \mathbf{R}_y = \begin{bmatrix} \cos\beta & 0 & \sin\beta \\ 0 & 1 & 0 \\ -\sin\beta & 0 & \cos\beta \end{bmatrix}$

c. $\beta = \gamma = 0$

Answer: $\beta = \gamma = 0$ and $\mathbf{R}_{zyx} = \mathbf{R}_z = \begin{bmatrix} 1 & 0 & 0 \\ 0 & \cos\alpha & -\sin\alpha \\ 0 & \sin\alpha & \cos\alpha \end{bmatrix}$

8.12. Find the equation of the following curves after the rotation transformation $x' = (x-y)/\sqrt{2}$, $y' = (x+y)/\sqrt{2}$:

a. $y = 0$ Answer: $x' = y'$

b. $y = x$ Answer: $x' = 0$

c. $y = x + 3$ Answer: $x' = -3/\sqrt{2}$

d. $x = 3$ Answer: $x' + y' = 3/\sqrt{2}$

e. $x^2 + y^2 = 1$ Answer: $x'^2 + y'^2 = 1$

f. $xy = 1$ Answer: $-x'^2 + y'^2 = 2$

g. $(x - 1)^2 + y^2 = 2$ Answer: $x'^2 + y'^2 - \sqrt{2}(x' + y') = 1$

h. $y = x^2$ Answer: $(x' + y')^2 = \sqrt{2}(-x' + y')$

i. Find the angle of rotation, ϕ. Answer: $\phi = 45°$

8.13. Find the rotation matrix for very small angles of rotation. Do this for two- and three-dimensional rotations.

Answer: For rotations about the origin in the xy plane, we have

$$\mathrm{R} = \begin{bmatrix} \cos\phi & -\sin\phi \\ \sin\phi & \cos\phi \end{bmatrix}.$$

If ϕ is in radians, then as $\phi \to 0$ (i.e., becomes very small), $\cos\phi \to 1$ and $\sin\phi \to \phi$. Thus, for small angles

$$\mathrm{R} = \begin{bmatrix} 1 & -\phi \\ \phi & 1 \end{bmatrix}.$$

Verify this by computing the determinant $\begin{vmatrix} 1 & -\phi \\ \phi & 1 \end{vmatrix}$, which is $1 + \phi^2$, where $\phi^2 \to 0$.

8.14. Show that the determinant of every translation transformation in homogeneous coordinates is equal to 1.

Answer: Compute the determinant of the matrix in the equation. Doing this, it is obvious that

$$\begin{vmatrix} 1 & 0 & 0 & t_x \\ 0 & 1 & 0 & t_y \\ 0 & 0 & 1 & t_z \\ 0 & 0 & 0 & 1 \end{vmatrix} = 1.$$

8.15. Show that the sum of the squares of any row or column of a proper rotation matrix is equal to one. This is the distinguishing characrteristic of a normalized matrix.

Answer:

Here is a simple example: Given the rotation matrix $\mathbf{R} = \begin{bmatrix} \cos\phi & -\sin\phi \\ \sin\phi & \cos\phi \end{bmatrix}$,

we see that $\cos^2\phi + \sin^2\phi = 1$ and similarly for other rows and columns.

8.16. Show that the scalar product of any pair of rows (or columns) of a proper rotation matrix is zero. A matrix with this property is orthogonal.

Answer:

As above we use the example of the rotation matrix $\mathbf{R} = \begin{bmatrix} \cos\phi & -\sin\phi \\ \sin\phi & \cos\phi \end{bmatrix}$

and treat each row (or column) as containing the components of a vector and peform the scalar product, producing zero.

Chapter 9 More Transformations

9.1. Given a homogeneous transformation H, partition this matrix and show that the 3x3 submatrix in the upper-left partition is always orthogonal.

Answer: The answer comes from analyzing the properties of matrix multiplication. Given any sequence of rotations expressed as homogeneous transformations, the net resultant is another rotation and thus orthogonal. For example, given two rotations $\mathbf{R}_1$ and $\mathbf{R}_2$:

$$\left[\begin{array}{ccc|c} & & & 0 \\ & \mathbf{R}_2 & & 0 \\ & & & 0 \\ \hline 0 & 0 & 0 & 1 \end{array}\right]\left[\begin{array}{ccc|c} & & & 0 \\ & \mathbf{R}_1 & & 0 \\ & & & 0 \\ \hline 0 & 0 & 0 & 1 \end{array}\right] = \left[\begin{array}{ccc|c} & & & 0 \\ & \mathbf{R}_3 & & 0 \\ & & & 0 \\ \hline 0 & 0 & 0 & 1 \end{array}\right],$$

where $\mathbf{R}_3 = \mathbf{R}_2\mathbf{R}_1$ and $|\mathbf{R}_3| = 1$. In addition, given any combination of a rotation and translation, the net resultant is the rotation and some translation. For example,

$$\left[\begin{array}{ccc|c} & & & 0 \\ & \mathbf{R}_1 & & 0 \\ & & & 0 \\ \hline 0 & 0 & 0 & 1 \end{array}\right]\left[\begin{array}{ccc|c} 1 & 0 & 0 & t_x \\ 0 & 1 & 0 & t_y \\ 0 & 0 & 1 & t_z \\ \hline 0 & 0 & 0 & 1 \end{array}\right] = \left[\begin{array}{ccc|c} & & & t_{14} \\ & \mathbf{R}_1 & & t_{24} \\ & & & t_{34} \\ \hline 0 & 0 & 0 & 1 \end{array}\right],$$

where the t_{i4} are expressions containing r_{ij} and t_x, t_y, t_z terms. The presence of the zeros in the last row and column of the homogeneous transformation matrix describing the rotation $\mathbf{R}_1$, and the presence of 1's on the diagonal and 0's elsewhere in the 3x3 submatrix ensures and yields $\mathbf{R}_1$ in the 3x3 submatrix of the resultant homogeneous transformation matrix.

9.2. Show that the determinant of the translation transformation in homogeneous coordinates is equal to 1.

Answer: Compute the determinant to prove that

$$\begin{vmatrix} 1 & 0 & 0 & t_x \\ 0 & 1 & 0 & t_y \\ 0 & 0 & 1 & t_z \\ 0 & 0 & 0 & 1 \end{vmatrix} = 1.$$

9.3. Sketch the central inversion of a rectangle in the plane, and discuss the orientation of the original and inverted rectangle.

Answer: Orientation is preserved. This inversion produces the same image as a rotation of 180° (Figure 9.38), and in both images the order of the vertices ABCD is counterclockwise.

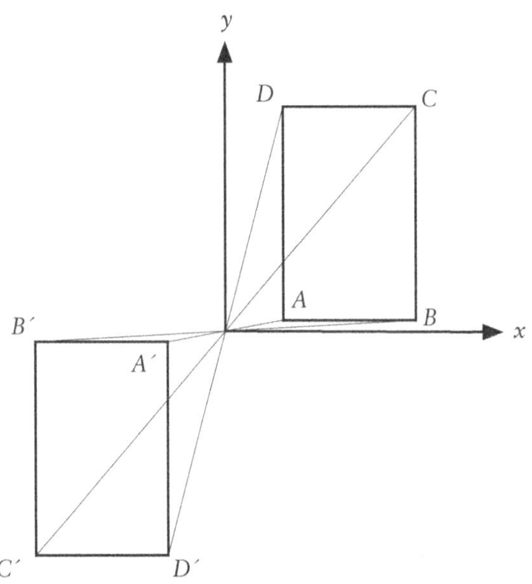

Figure 9.38 Inversion of a rectangle in the origin

9.4. Find the matrix and Cartesian equations that describe a reflection fixing the line $x = d$ parallel to the y axis.

Answer: $\mathbf{p}' = \begin{bmatrix} -1 & 0 & 2d \\ 0 & 1 & 0 \\ 0 & 0 & 1 \end{bmatrix} \mathbf{p}$ and $\begin{cases} x' = -x + 2d \\ y' = y \end{cases}$

9.5. Find the reflection matrix that maps any point on the plane onto its mirror image across the line $y = x / \sqrt{3}$.

Answer: The line $y = x / \sqrt{3}$ passes through the origin at an angle of 30° with the x axis. Thus,

$$\mathbf{p}' = \begin{bmatrix} \cos 30° & -\sin 30° & 0 \\ \sin 30° & \cos 30° & 0 \\ 0 & 0 & 1 \end{bmatrix} \begin{bmatrix} 1 & 0 & 0 \\ 0 & -1 & 0 \\ 0 & 0 & 1 \end{bmatrix} \begin{bmatrix} \cos 30° & \sin 30° & 0 \\ -\sin 30° & \cos 30° & 0 \\ 0 & 0 & 1 \end{bmatrix} \mathbf{p}$$

$$\cdots = \begin{bmatrix} \sqrt{3}/2 & -1/2 & 0 \\ 1/2 & \sqrt{3}/2 & 0 \\ 0 & 0 & 1 \end{bmatrix} \begin{bmatrix} 1 & 0 & 0 \\ 0 & -1 & 0 \\ 0 & 0 & 1 \end{bmatrix} \begin{bmatrix} \sqrt{3}/2 & 1/2 & 0 \\ -1/2 & \sqrt{3}/2 & 0 \\ 0 & 0 & 1 \end{bmatrix} \mathbf{p}$$

$$\cdots = \begin{bmatrix} 1/2 & \sqrt{3}/2 & 0 \\ \sqrt{3}/2 & -1/2 & 0 \\ 0 & 0 & 1 \end{bmatrix} \mathbf{p}$$

9.6. Show that the product of the reflection matrices $\begin{bmatrix} 0 & 1 & 0 \\ 1 & 0 & 0 \\ 0 & 0 & 1 \end{bmatrix}$ and $\begin{bmatrix} 0 & -1 & 0 \\ -1 & 0 & 0 \\ 0 & 0 & 1 \end{bmatrix}$ is a rotation. Describe the rotation.

Answer: $\begin{bmatrix} 0 & 1 & 0 \\ 1 & 0 & 0 \\ 0 & 0 & 1 \end{bmatrix} \begin{bmatrix} 0 & -1 & 0 \\ -1 & 0 & 0 \\ 0 & 0 & 1 \end{bmatrix} = \begin{bmatrix} -1 & 0 & 0 \\ 0 & -1 & 0 \\ 0 & 0 & 1 \end{bmatrix}$

and

$$\det \begin{bmatrix} -1 & 0 & 0 \\ 0 & -1 & 0 \\ 0 & 0 & 1 \end{bmatrix} = +1$$

Therefore, it is a rotation of 180° about the origin.

9.7. Show that any polygon transforms into a similar polygon under a scaling transformation.

Answer: Divide the polygon into triangular sectors and use vectors to represent the edges. Corresponding edges are parallel, since $\mathbf{p}_j - \mathbf{p}_i$ is parallel to $k\,(p_j - p_i)$. Only the length has changed (Figure 9.39).

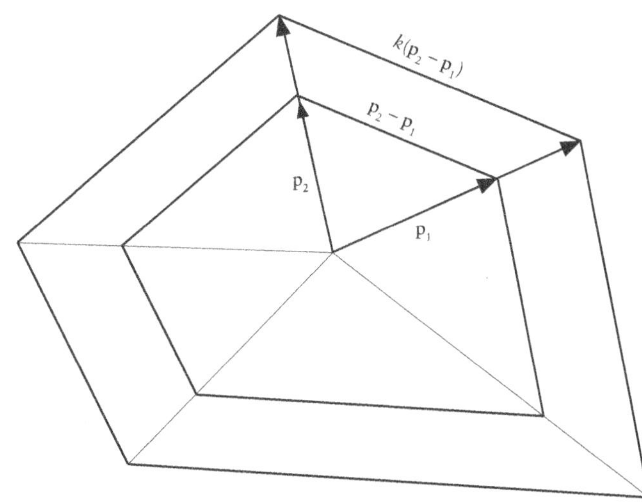

Figure 9.39 Scaling and similar polygons

9.8. Show that angle size is preserved under a scaling transformation s.
Answer: The angle θ between the two vectors **p** and **q** before scaling is

$\theta = \cos^{-1}\dfrac{\mathbf{p}\bullet\mathbf{q}}{|\mathbf{p}||\mathbf{q}|}$, and after dilation is

$$\theta' = \cos^{-1}\frac{\mathbf{p}'\bullet\mathbf{q}'}{|\mathbf{p}'||\mathbf{q}'|} = \cos^{-1}\frac{s\mathbf{p}\bullet s\mathbf{q}}{|s\mathbf{p}||s\mathbf{q}|} = \cos^{-1}\frac{s^2(\mathbf{p}\bullet\mathbf{q})}{s^2|\mathbf{p}||\mathbf{q}|} = \cos^{-1}\frac{(\mathbf{p}\bullet\mathbf{q})}{|\mathbf{p}||\mathbf{q}|},$$

or $\theta' = \theta$.

9.9. Find the center of scaling for the transformation given by the matrix

$$\begin{bmatrix} s & 0 & c_1 \\ 0 & s & c_2 \\ 0 & 0 & 1 \end{bmatrix}.$$

Answer: $x_c = \dfrac{c_1}{1-s}, \quad y_c = \dfrac{c_2}{1-s}$

9.10. Show that $\mathbf{S}\mathbf{S}^{-1} = 1$.

Answer: $$\begin{bmatrix} s & 0 & 0 & 0 \\ 0 & s & 0 & 0 \\ 0 & 0 & s & 0 \\ 0 & 0 & 0 & 1 \end{bmatrix}\begin{bmatrix} 1/s & 0 & 0 & 0 \\ 0 & 1/s & 0 & 0 \\ 0 & 0 & 1/s & 0 \\ 0 & 0 & 0 & 1 \end{bmatrix} = \begin{bmatrix} 1 & 0 & 0 & 0 \\ 0 & 1 & 0 & 0 \\ 0 & 0 & 1 & 0 \\ 0 & 0 & 0 & 1 \end{bmatrix}$$

9.11. Write the Cartesian equations for a shear transformation in the plane that fixes the y axis.
Answer: $x' = x, \quad y' = k_x x + y$

Chapter 10 Vector-Defined Geometric Objects I

10.1. Given $\mathbf{a} = [\,6 \;\; -1 \;\; -2\,]$, $\mathbf{b} = [\,3 \;\; 2 \;\; 4\,]$, and $\mathbf{c} = [\,7 \;\; 0 \;\; 2\,]$, write the vector equation of a line

a. Through **a** and parallel to **b** — Answer: $\mathbf{p} = \mathbf{a} + u\mathbf{b}$

b. Through **b** and parallel to **c** — Answer: $\mathbf{p} = \mathbf{b} + u\mathbf{c}$

c. Through **c** and parallel to **a** — Answer: $\mathbf{p} = \mathbf{c} + u\mathbf{a}$

d. Through **a** and parallel to **a** — Answer: $\mathbf{p} = \mathbf{a} + u\mathbf{a} = (u + 1)\mathbf{a}$

e. Through **b** and parallel to **a** — Answer: $\mathbf{p} = \mathbf{b} + u\mathbf{a}$

10.2. Find the equations of the x, y, and z vector components for the line segments given by the following pairs of endpoints:

a. $\mathbf{p}_0 = [\ 0\ \ 0\ \ 0\ ]$, $\mathbf{p}_1 = [\ 1\ \ 1\ \ 1\ ]$

Answer: $x = u$, $y = u$, $z = u$

b. $\mathbf{p}_0 = [\ -3\ \ 1\ \ 6\ ]$, $\mathbf{p}_1 = [\ 2\ \ 0\ \ 7\ ]$

Answer: $x = -3 + 5u$, $y = 1 - u$, $z = 6 + u$

c. $\mathbf{p}_0 = [\ 1\ \ 1\ \ -4\ ]$, $\mathbf{p}_1 = [\ 5\ \ -3\ \ 9\ ]$

Answer: $x = 1 + 4u$, $y = 1 - 4u$, $z = -4 + 13u$

d. $\mathbf{p}_0 = [\ 6\ \ 8\ \ 8\ ]$, $\mathbf{p}_1 = [\ -10\ \ 0\ \ -3\ ]$

Answer: $x = 6 - 16u$, $y = 8 - 8u$, $z = 8 - 11u$

e. $\mathbf{p}_0 = [\ 0\ \ 0\ \ 1\ ]$, $\mathbf{p}_1 = [\ 0\ \ 0\ \ -1\ ]$

Answer: $x = 0$, $y = 0$, $z = 1 - 2u$

An interesting variation of this vector geometry of a straight line will help us find the vector equation of a line through two given points, say $\mathbf{p}_0$ and $\mathbf{p}_1$. We begin by examining the following equation (a sketch of these vectors might help):

$$\mathbf{p} = \mathbf{p}_0 + u(\mathbf{p}_1 - \mathbf{p}_0).$$

If we restrict the range of values of u to the interval $0 \le u \le 1$, then this equation defines a line segment extending from $\mathbf{p}_0$ to $\mathbf{p}_1$. In algebraic form, this vector equation expands to

$$x = x_0 + u(x_1 - x_0),$$
$$y = y_0 + u(y_1 - y_0),$$
$$z = z_0 + u(z_1 - z_0).$$

10.3. Given $x = 3 + 2u$, $y = -6 + u$, and $z = 4$, find $\mathbf{p}_0$ and $\mathbf{p}_1$, with $0 \le u \le 1$.

Answer: $\mathbf{p}_0 = [\ 3\ \ -6\ \ 4\ ]$, $\mathbf{p}_1 = [\ 5\ \ -5\ \ 4\ ]$

Notice that this line lies in the $z = 4$ plane, so that $z_0 = 4$ and $z_1 = 4$. To find x_0 and y_0, set $u = 0$. To find x_1 and y_1, set $u = 1$.

10.4. Describe the difference between the following line segments:

For line 1, $\mathbf{p}_0 = [\ 2\ \ 1\ \ -2\ ]$ and $\mathbf{p}_1 = [\ 3\ \ -3\ \ 1\ ]$.

For line 2, $\mathbf{p}_0 = [\ 3\ \ -3\ \ 1\ ]$ and $\mathbf{p}_1 = [\ 2\ \ 1\ \ -2\ ]$.

Answer: The line segments are identical, with opposite directions of parameterization.

10.5. Write the vector equation of the plane passing through **a** and parallel to **b** and **c**.

Answer: $\mathbf{p} = \mathbf{a} + u\mathbf{b} + w\mathbf{c}$

There are four common ways to define a plane in three dimensions using vector equations. One way is by the vector equation of a plane through $\mathbf{p}_0$ and parallel to two independent vectors **s** and **t**:

$$\mathbf{p} = \mathbf{p}_0 + u\mathbf{s} + w\mathbf{t},$$

where $\mathbf{s} \neq k\mathbf{t}$ and where u and w are scalar independent variables multiplying $\mathbf{s}$ and $\mathbf{t}$, respectively. The vector $\mathbf{p}$ represents the set of points defining a plane as the parameters u and w vary independently.

10.6. Write the vector equation of a plane that passes through the origin and is perpendicular to the y axis.

Answer: $\mathbf{p} = \mathbf{p}_0 + u\mathbf{s} + w\mathbf{t}$, where $\mathbf{p} = [\,x \;\; y \;\; z\,]$, $\mathbf{p}_0 = [\,0 \;\; 0 \;\; 0\,]$, $\mathbf{s} = [\,s_x \;\; 0 \;\; s_z\,]$, and $\mathbf{t} = [\,t_x \;\; 0 \;\; t_z\,]$ and where s_x, s_y, t_x, and t_z are arbitrary real-number constants.

10.7. Show algebraically that the equation $\mathbf{p}_0 \bullet \mathbf{n} = d$ represents the equation of a plane passing through $\mathbf{p}_0$, normal to the unit vector $\mathbf{n}$, where d is the perpendicular distance from the origin to the plane. Also, interpret this equation graphically with an appropriate sketch.

Answer: $(d\mathbf{n} - \mathbf{p}_0) \bullet d\mathbf{n} = 0$; and for any point $\mathbf{p}$, $(\mathbf{p} - \mathbf{p}_0) \bullet \mathbf{n} = 0$.

10.8. Show that the intersection of three planes is given by

$$\mathbf{p} = \frac{d_1(\mathbf{n}_2 \times \mathbf{n}_3) + d_2(\mathbf{n}_3 \times \mathbf{n}_1) + d_3(\mathbf{n}_1 \times \mathbf{n}_2)}{\mathbf{n}_1 \bullet \mathbf{n}_2 \times \mathbf{n}_3}.$$

Answer: This equation is verified by showing that it satisfies the equation for each of the three planes: $\mathbf{p} \bullet \mathbf{n}_1 = d_1$, $\mathbf{p} \bullet \mathbf{n}_2 = d_2$, and $\mathbf{p} \bullet \mathbf{n}_3 = d_3$.

10.9. Compute the distance between each of the following pairs of points:

a. (–2.7, 6.5, 0.8), (5.1, –5.7, 1.9) Answer: 14.522

b. (1, 1, 0), (4, 6, –3) Answer: 6.557

c. (7, –4, 2), (0, 2.7, –0.3) Answer: 9.959

d. (–3, 0, 0), (7, 0, 0) Answer: 10.00

e. (10, 9, –1), (3, 8, 3) Answer: 8.124

Use the observation that the minimum distance between two arbitrary points in space $\mathbf{p}_1$ and $\mathbf{p}_2$ is

$$d_{\min} = |\,\mathbf{p}_2 - \mathbf{p}_1\,| = \sqrt{(\mathbf{p}_2 - \mathbf{p}_1) \bullet (\mathbf{p}_2 - \mathbf{p}_1)},$$

or

$$d = \sqrt{(x_2 - x_1)^2 + (y_2 - y_1)^2 + (z_2 - z_1)^2}.$$

10.10. Compute the coordinates of the points in Exercise 10.9 relative to a coordinate system centered at (3, –1, 0) in the original system and parallel to it.

Answer: a. (–5.7, 7.5, 0.8) and (2.1, –4.7, 1.9)

Answer: b. (–2, 2, 0) and (1, 7, –3)

Answer: c. (4, –3, 2) and (–3, 3.7, –0.3)

Answer: d. (–6, 1, 0) and (4, 1, 0)

Answer: e. (7, 10, –1) and (0, 9, 3)

10.11. Compute the distance between each of the points found for Exercise 10.10.

a. (–5.7, 7.5, 0.8) and (2.1, –4.7, 1.9) Answer: 14.52

b. (–2, 2, 0) and (1, 7, –3) Answer: 6.557

c. (4, –3, 2) and (–3, 3.7, –0.3) Answer: 9.959

d. (–6, 1, 0) and (4, 1, 0) Answer: 10.000

e. (7, 10, –1) and (0, 9, 0) Answer: 8.124

10.12. Show that the distance between any pair of points is independent of the coordinate system chosen.

Answer: If one coordinate system is displaced with respect to another one, then $x' = x - \Delta x$, $y' = y - \Delta y$, and $z' = z - \Delta z$, where Δx, Δy, and Δz are the displacements. Given two points in the original system, (x_1, y_1, z_1) and (x_2, y_2, z_2), the distance between them is

$$\sqrt{(x_2 - x_1)^2 + (y_2 - y_1)^2 + (z_2 - z_1)^2}$$

The coordinates of points in the displaced system are $(x_1 - \Delta x,\ y_1 - \Delta y,\ z_1 - \Delta z)$ and $(x_2 - \Delta x, y_2 - \Delta y, z_2 - \Delta z)$ and the distance between them is

$$\sqrt{[(x_2 - \Delta x) - (x_1 - \Delta x)]^2 + [(y_2 - \Delta y) - (y_1 - \Delta y)]^2 + [(z_2 - \Delta z) - (z_1 - \Delta z)]^2},$$

which reduces to the original distance formula, since the Δx, Δy, and Δz terms cancel.

10.13. Compute the midpoint between the pairs of points given in Exercise 10.9.

a. (–2.7, 6.5, 0.8), (5.1, –5.7, 1.9) Answer: (1.2, 0.4, 1.43)

b. (1, 1, 0), (4, 6, –3) Answer: (2.5, 3.5, –1.5)

c. (7, –4, 2), (0, 2.7, –0.3) Answer: (3.5, –0.65, 0.85)

d. (–3, 0, 0), (7, 0, 0) Answer: (2.0, 0.0, 0.0)

e. (10, 9, –1), (3, 8, 3) Answer: (6.5, 8.5, 1.0)

10.14. Given an arbitrary set of points, find the coordinates of the vertices of a rectangular box that just encloses it.

Answer: Find the minimum and maximum x, y, and z values in the point set and from these construct the coordinates of the corners of the enclosing box, obtaining:

(1) $(x_{min}, y_{min}, z_{min})$

(2) $(x_{max}, y_{min}, z_{min})$

(3) $(x_{max}, y_{max}, z_{min})$

(4) $(x_{min}, y_{max}, z_{min})$

(5) $(x_{min}, y_{min}, z_{max})$

(6) $(x_{max}, y_{min}, z_{max})$

(7) $(x_{max}, y_{max}, z_{max})$

(8) $(x_{min}, y_{max}, z_{max})$

10.15. Find the coordinates of the eight vertices of a rectangular box that just encloses the ten points given in Exercise 10.9.

Answer:

(1) $(x_{min}, y_{min}, z_{min}) = (-3, -5.7, -3)$

(2) $(x_{max}, y_{min}, z_{min}) = (10, -5.7, -3)$

(3) $(x_{max}, y_{max}, z_{min}) = (10, 9, -3)$

(4) $(x_{min}, y_{max}, z_{min}) = (-3, 9, -3)$

(5) $(x_{min}, y_{min}, z_{max}) = (-3, -5.7, 3)$

(6) $(x_{max}, y_{min}, z_{max}) = (10, -5.7, 3)$

(7) $(x_{max}, y_{max}, z_{max}) = (10, 9, 3)$

(8) $(x_{min}, y_{max}, z_{max},) = (-3, 9, 3)$

10.16. Given that Δ_i is a constant for all $\mathbf{p}_i$, that is, $\Delta_i = (\Delta x_i, \Delta y_i) = (\Delta x, \Delta y)$, find $\mathbf{p}_4$ in terms of $\mathbf{p}_0$ and Δ_i.

Answer: Compute each point sequentially from $\mathbf{p}_0$ to $\mathbf{p}_4$:

$$\mathbf{p}_0 = (x_0, y_0)$$

$$\mathbf{p}_1 = (x_0 + \Delta x,\ y_0 + \Delta y)$$

$$\mathbf{p}_2 = (x_0 + 2\Delta x,\ y_0 + 2\Delta y)$$

$$\mathbf{p}_3 = (x_0 + 3\Delta x,\ y_0 + 3\Delta y)$$

$$\mathbf{p}_4 = (x_0 + 4\Delta x,\ y_0 + 4\Delta y)$$

We define the coordinates of each relative point with reference to the coordinates of the point preceding it.

10.17. Find the set of Δ_i's for the vertex points of a square whose sides are three units long and with $\mathbf{p}_0 = (1,0)$. Assume that the sides of the square are parallel to the x, y coordinate axes, and proceed counterclockwise.

Answer: $\Delta_1 = (\Delta x_1 = 3,\ \Delta y_1 = 0)$, $\Delta_2 = (\Delta x_2 = 0,\ \Delta y_2 = 3)$, $\Delta_3 = (\Delta x_3 = -3,\ \Delta y_3 = 0)$

10.18. Repeat Exercise 10.17 for a square whose sides are four units long and with $\mathbf{p}_0 = (-2, -2)$.

Answer: $\Delta_1 = (\Delta x_1 = 4, \Delta y_1 = 0)$, $\Delta_2 = (\Delta x_2 = 0, \Delta y_2 = 4)$, $\Delta_3 = (\Delta x_3 = -4, \Delta y_3 = 0)$

10.19. Repeat Exercise 10.17 for $\mathbf{p}_0 = (1, -4)$.

Answer: $\Delta_1 = (\Delta x_1 = 4, \Delta y_1 = 0)$, $\Delta_2 = (\Delta x_2 = 0, \Delta y_2 = 4)$, $\Delta_3 = (\Delta x_3 = -4, \Delta y_3 = 0)$

10.20. Show that the line joining the midpoints of two sides of a triangle is parallel to the third side and has one-half its magnitude.

Answer: Let **a** and **b** be two sides of a triangle. Then **b** – **a** is the third side. The midpoints of **a** and **b** are 0.5**a** and 0.5**b**, respectively. A line joining these midpoints is represented by the vector 0.5 (**b** – **a**), which is parallel to and half the length of **b** – **a**.

10.21. Find the midpoint of the line segment between $\mathbf{p}_0 = [\ 3 \quad 5 \quad 1\ ]$ and $\mathbf{p}_1 = [\ -2 \quad 6 \quad 4\ ]$.

Answer: The midpoint is $0.5(\mathbf{p}_0 + \mathbf{p}_1) = [\ 0.5 \quad 5.5 \quad 2.5\ ]$.

10.22. Write the vector equation of a line through two points $\mathbf{p}_0$ and $\mathbf{p}_1$.

Answer: $\mathbf{r} = \mathbf{r}_0 + u(\mathbf{r} - \mathbf{r}_0)$, where u is a real number.

10.23. Write the vector equation of a plane containing the three noncollinear points $\mathbf{p}_0$, $\mathbf{p}_1$, and $\mathbf{p}_2$.

Answer: $\mathbf{r} = \mathbf{r}_0 + u(\mathbf{r}_1 - \mathbf{r}_0) + w(\mathbf{r}_2 - \mathbf{r}_1)$, where u and w are real numbers.

Index

www.ingramcontent.com/pod-product-compliance
Lightning Source LLC
Chambersburg PA
CBHW061356210326
41598CB00035B/5998

* 9 7 8 0 8 3 1 1 3 6 5 5 0 *